Oil's Industry History

Oil's Industry History

A Deep Dive

Sam Loray

UNIEK ENTERPRISES

CONTENTS

INDEX

Introduction

The historical backdrop of the oil business is an intriguing excursion that traverses hundreds of years and landmasses, a story of human creativity, asset double-dealing, and international elements that have formed the cutting edge world. From the unassuming starting points of oil extraction in old civic establishments to the rambling worldwide organization of investigation, creation, and circulation that we see today, the oil business plays had a critical impact in driving monetary advancement, powering mechanical progressions, and, on occasion, igniting international strains and ecological difficulties. In this profound jump into the historical backdrop of the oil business, we will investigate the key achievements, people, and occasions that have characterized this mind boggling and diverse industry.

The foundations of the oil business can be followed back to vestige, where early civic establishments outfit the force of oil based substances for various purposes. The old Sumerians, for instance, utilized bitumen, a normally happening type of petrol, as a structure material and for waterproofing. Also, the antiquated Egyptians used bitumen in preservation rehearses and for fixing their boats. These early purposes of oil based commodities give a brief look into the old world's restricted comprehension of the huge likely secret underneath the World's surface.

Nonetheless, it was only after the late nineteenth century that the genuine size of the oil business' true capacity started to arise. The fast industrialization and innovative progressions of the nineteenth century provoked an unquenchable interest for energy sources to drive apparatus, fuel transportation, and enlighten urban communities. It was during this period that the oil business made its most memorable groundbreaking strides towards turning into a worldwide monetary force to be reckoned with.

Quite possibly of the most urgent crossroads throughout the entire existence of the oil business happened during the nineteenth century when Colonel Edwin Drake effectively bored the world's most memorable business oil well in Titusville, Pennsylvania, in 1859. This occasion denoted the introduction of the advanced oil industry and introduced a time of phenomenal development and advancement. Drake's revelation prompted the business creation of oil, a product that would before long change the worldwide energy scene.

The late nineteenth century additionally saw the ascent of John D. Rockefeller and his Standard Oil Organization. Rockefeller's essential vision and heartless business strategies permitted him to unite command over numerous parts of the oil business, from boring to refining to dissemination. Standard Oil's predominance was outright to the point that it prompted antitrust activities and the possible separation of the organization in 1911. Notwithstanding its questionable heritage, Standard Oil's inheritance remains profoundly imbued throughout the entire existence of the oil business.

The oil business' development was not restricted to the US. All through the late nineteenth and mid twentieth hundreds of years, critical oil revelations were made in different regions of the planet, from the oilfields of Baku in Azerbaijan to the immense stores in the Center East. The improvement of pipelines and foundation to move oil to far off business sectors worked with its worldwide circulation, making it a basic asset for industrialized countries.

As the oil business extended, it significantly affected society and the worldwide economy. The improvement of the gas powered motor changed transportation and made ready for the cutting edge car. The reception of oil as an essential wellspring of energy considered the large scale manufacturing of vehicles, prompting the fast development of the car business and a change in how individuals lived and functioned.

The oil business' development likewise changed international affairs. The essential significance of oil holds prompted extraordinary rivalry and clashes over admittance to these important assets. The two universal conflicts of the twentieth century were to some degree formed by the longing to get oil-rich locales, like the Center East. The revelation of tremendous oil holds in nations like Saudi Arabia and Iran further featured the international meaning of oil, as countries competed for control and impact in the locale.

The oil business' set of experiences isn't without its dim parts, in any case. Ecological worries started to arise as the business' impression extended. Oil slicks, air contamination, and other natural fiascos brought issues to light about the business' environmental effect. Endeavors to resolve these issues, like the foundation of natural guidelines and the advancement of cleaner innovations, have been continuous difficulties for the business.

All through the twentieth 100 years, the oil business kept on developing, set apart by mechanical progressions and advancements. Seaward boring, for instance, permitted admittance to tremendous stores underneath the sea depths, while the improvement of profound penetrating strategies empowered the extraction of oil from beforehand unavailable locales. These innovative leap forwards extended the business' ability and expanded its geographic reach.

The last 50% of the twentieth century carried another aspect to the oil business - the worldwide energy emergency. The 1970s saw a progression of oil shocks, driven by international pressures in the Center East and rising worldwide interest for oil. These shocks uncovered the world's weakness to oil cost variances and set off endeavors to diminish reliance on petroleum products through energy preservation and the improvement of elective energy sources.

As we entered the 21st 100 years, the oil business confronted an intersection of difficulties, including worries about environmental change and the exhaustion of effectively open oil saves. These provokes prompted a developing accentuation on supportability and environmentally friendly power sources.

While oil stays a predominant energy source, the business' drawn out future is being reshaped by a shift towards cleaner and all the more harmless to the ecosystem choices.

The historical backdrop of the oil business is deficient without recognizing the commitments and effects of key figures and associations. From people like John D. Rockefeller and J. Paul Getty, whose names are inseparable from oil magnates, to the associations that have characterized the business scene, like ExxonMobil, BP, and Shell, these players have made a permanent imprint on the world's energy elements.

The oil business' set of experiences is likewise interwoven with political and monetary variables. The development of the Association of the Oil Trading Nations (OPEC) in 1960 denoted a defining moment in the business' elements. OPEC's capacity to all in all control oil costs and creation levels has had extensive impacts on the worldwide economy and international affairs.

The historical backdrop of the oil business is packed with complex and frequently disputable occasions, for example, the Inlet War, the Deepwater Skyline oil slick, and discussions over the Cornerstone XL pipeline. These occurrences highlight the business' persevering through job in forming political talk, ecological approaches, and worldwide security concerns.

As of late, the oil business has confronted another arrangement of difficulties. The quick development of environmentally friendly power sources, the rise of electric vehicles, and the push for feasible practices are reshaping the business' future. The change to a low-carbon economy addresses an essential shift that the oil business should explore to stay pertinent and maintainable in the long haul.

The historical backdrop of the oil business is an account of development, desire, and change. It plays had a focal impact in the improvement of current culture, fueling our economies, associating our reality, and affecting our legislative issues. In any case, it has likewise left a huge natural impression and brought up basic issues about our energy future.

As we dig further into the oil business' set of experiences, we will look at the business' worldwide come to, its effect on social orders and economies, the developments that have driven its development, the difficulties it has confronted, and the continuous endeavors to adjust its part in a quickly impacting world. This profound jump into the oil business' set of experiences will reveal insight into the complexities, triumphs, contentions, and vulnerabilities that have characterized this fundamental area, and it will assist us with better figuring out the job of oil from before, present, and future.

1

Chapter 1

The Birth of an Industry

In the chronicles of mankind's set of experiences, there have been minutes that have always changed the direction of human progress. These minutes are not generally set apart by fabulous fights or political upheavals; in some cases, they are peaceful and honest, unfurling continuously over the long haul. The introduction of an industry is one such extraordinary occasion. A peculiarity has impacted social orders, economies, and societies all through the ages.

The introduction of an industry is an intricate and complex interaction. It is a combination of different elements, thoughts, and developments, all approaching together to make something new and uncommon. At its center, it addresses a principal shift in the manner in which individuals live, work, and communicate with their general surroundings.

The foundations of an industry's introduction to the world can frequently be followed back to a solitary thought or innovation. These impetuses of progress are much of the time the aftereffect of long periods of trial and error, logical revelation, and human creativity. They are the sparkles that touch off an upheaval, getting rolling a chain of occasions that will at last reshape the world.

Take, for instance, the introduction of the car business. Everything started with the advancement of the gas powered motor in the late nineteenth hundred years. This advanced innovation, which considered the effective change of fuel into mechanical energy, prepared for the production of the principal autos. It was a snapshot of significant importance, as it denoted the progress from horse-attracted carriages to self-moved vehicles.

The introduction of the auto business had broad outcomes. It changed transportation as well as significantly affected metropolitan preparation, framework improvement, and, surprisingly, the idea of work. Industrial facilities jumped up to deliver vehicles and related parts, making position and invigorating monetary development. Streets and interstates were worked to oblige the new method of transport, changing urban communities and scenes. The manner in which individuals lived and worked changed too, as driving became simpler, and the idea of rural areas arose.

This change of society was not one of a kind to the auto business. From the beginning of time, the introduction of different enterprises has significantly affected the world. The Modern Upheaval of the eighteenth and nineteenth hundreds of years saw the introduction of the material, coal, and iron businesses, among others. These businesses generally changed the idea of work, moving from agrarian and make based economies to modern and manufacturing plant based frameworks. The introduction of the message and the railroad business altered correspondence and transportation, associating far off places and empowering worldwide exchange.

For each situation, the introduction of an industry put into high gear an outpouring of changes that contacted each part of life. It was not just about the formation of new items or administrations; it was about a change in the actual structure holding the system together itself.

One of the characterizing elements of the introduction of an industry is the fast speed of advancement and improvement. In the beginning phases, new innovations and thoughts frequently arise in a divided and turbulent way. There is a feeling of trial and error, as people and organizations look to benefit from the arising open doors. This can prompt a time of serious rivalry and vulnerability as various players strive for predominance.

The introduction of the PC business, for instance, was portrayed by a multiplication of various equipment and programming frameworks. During the 1970s and 1980s, organizations like IBM, Apple, and Microsoft all assumed key parts in molding the business, each with its own vision and approach. It was a time of quick change and vulnerability, as the business battled to track down its balance.

In this climate, development is many times driven by a feeling of business and hazard taking. Business people and creators will take risks on doubtful thoughts, and financial speculators will put resources into new and untested endeavors. This soul of advancement permits businesses to develop and develop, as new players and thoughts enter the scene.

The introduction of an industry additionally frequently includes the union of various fields of information. It is when researchers, architects, and financial specialists meet up to pool their skill and assets. The introduction of the drug business, for example, depended on the cooperation of scientific experts, researcher, and clinical experts. Their joined endeavors prompted the advancement of life-saving medications and clinical medicines that have changed medical services.

The introduction of the web and the innovation business is one more convincing illustration of interdisciplinary cooperation. It united PC researchers, electrical specialists, and business visionaries to make a worldwide organization that has changed how data is shared and gotten to. The web has led to endless different ventures and has turned into a fundamental piece of current life.

As businesses mature, they frequently go through periods of combination and normalization. In the good 'ol days, there might be a wide assortment of contending items and guidelines. Be that as it may, over the long run, certain innovations and approaches become prevailing, prompting a more steady and unsurprising climate.

The normalization of the electrical business is a decent outline of this interaction. In the late nineteenth 100 years, there were different contending frameworks for producing and dispersing power. Thomas Edison supported direct current (DC), while George Westinghouse and Nikola Tesla pushed for substituting current (AC). The "Battle of Flows" seethed on, with the two sides promoting the upsides of their particular frameworks.

Eventually, AC won as the prevailing norm for electrical power dissemination because of its productivity over significant distances. This normalization considered the boundless sending of electrical matrices and the zap of urban communities, preparing for present day electrical machines and ventures.

The introduction of an industry likewise significantly affects instruction and the advancement of particular information. As new businesses arise, there is a developing interest for people with mastery in the pertinent fields. This frequently prompts the foundation of instructive establishments and preparing programs devoted to these enterprises.

The introduction of the airplane business, for instance, provoked the making of aeronautical designing projects at colleges. As the business created, it required a labor force with specific abilities in regions like streamlined features, materials science, and drive frameworks. These instructive drives upheld the development of the business as well as added to progressions in logical information.

In addition, the introduction of an industry every now and again prompts the advancement of expert affiliations and guidelines associations. These elements assume a pivotal part in laying out prescribed procedures, setting wellbeing guidelines, and cultivating joint effort inside the business. For example, the introduction of the clinical gadget industry has brought about associations like the U.S. Food and Medication Organization (FDA), which controls and screens the wellbeing and adequacy of clinical gadgets to safeguard general wellbeing.

Notwithstanding instruction and guidelines, the introduction of an industry frequently achieves huge legitimate and administrative changes. State run administrations and policymakers should adjust to the rise of new innovations and enterprises, creating regulations and guidelines to guarantee public wellbeing and fair contest.

The introduction of the media communications industry, for example, required the production of administrative bodies like the Government Correspondences Commission (FCC) in the US. These offices are liable for allotting radio frequencies, upholding broadcast communications regulations, and guaranteeing that purchasers approach reasonable and dependable correspondence administrations.

The introduction of media outlets, including film and TV, likewise provoked the improvement of copyright and licensed innovation regulations. These legitimate structures safeguard the imaginative works of people and associations, encouraging advancement and creative articulation while guaranteeing that makers are made up for their commitments.

Besides, the introduction of an industry frequently prompts financial moves and changes in the work market. New ventures can make occupations and animate financial development, yet they can likewise upset existing areas. The ascent of web based business and online retail, for instance, has changed the retail business and production network coordinated operations, affecting conventional physical stores and setting out new open doors in the advanced domain.

Globalization is one more outcome of the introduction of ventures. As new innovations and businesses arise, they frequently rise above public limits and cultivate worldwide cooperation. The aviation and car enterprises, for example, have prompted associations among organizations and states from various nations, driving innovative headways and empowering the investigation of room and the improvement of world-wide exchange organizations.

The introduction of the drug business has additionally prodded global participation in the field of general wellbeing. State run administrations, research establishments, and drug organizations from around the world have cooperated to foster immunizations and meds that battle illnesses and work on human prosperity.

Notwithstanding globalization, the introduction of an industry can have international ramifications. The interest for unrefined components, energy assets, and intriguing earth components can prompt contest and clashes among countries. The energy business, for instance, has frequently been at the focal point of international strains as nations look to get their energy supplies and safeguard their inclinations.

1.1 Introduction to the book's theme.

The subject of this book is a significant investigation of the human condition, an endeavor to comprehend and get a handle on the intricacies that characterize our reality. It dives profound into the core of being human, the inquiries that have consumed the personalities of logicians, researchers, and craftsmen over the entire course of time.

At its center, this book is a challenge to leave on an excursion of contemplation, an excursion that looks to disentangle the mind boggling embroidery of human encounters, feelings, and yearnings. It takes us on a mission to find the consistent ideas that tight spot all of us, in spite of our singular distinctions, and to examine the significance and reason for our reality in this immense and steadily evolving universe.

The investigation of the human condition is a deep rooted try, one that has been sought after through different disciplines and mediums. From writing to reasoning, from brain science to social science, the journey to comprehend the human experience has driven incalculable masterminds, researchers, and narrators to look for bits of knowledge and shrewdness.

Since forever ago, humankind has wrestled with central inquiries concerning our spot on the planet. Questions like: What is the idea of cognizance? How would we track down significance in an apparently unconcerned universe? What drives our longings, fears, and inspirations? What is the wellspring of our innovativeness and the wellspring of our feelings?

This book doesn't really give conclusive solutions to these inquiries. All things considered, it tries to cultivate a more profound appreciation and reflection upon them. It urges perusers to mull over their own encounters and points of view and take part in an exchange with the extraordinary scholars and craftsmen who have contemplated these equivalent inquiries all through the ages.

One of the fundamental parts of investigating the human condition is the acknowledgment of our common mankind. We as a whole, no matter what our experiences or convictions, face a typical arrangement of difficulties, delights, and distresses. The human experience is an embroidery woven from the strings of euphoria, distress, love, misfortune, trust, and gloom. It is a demonstration of the strength of the human soul, our ability for empathy, and our interminable quest for understanding.

All through the pages of this book, we will experience stories, bits of knowledge, and reflections that shed light on different aspects of the human condition. We will dig into the domain of feelings, investigating the range of human sentiments from affection and satisfaction to dread and anguish. We will wrestle with the idea of time and mortality, looking to comprehend how our attention to our limited presence shapes our lives and decisions.

The human experience is additionally profoundly interwoven with our ability for inventiveness and creative mind. We will investigate the innovative soul that has enlivened craftsmanship, writing, and logical disclosure. We will think about the manners by which narrating and account shape how we might interpret the world and ourselves.

In our journey to comprehend the human condition, we should likewise deal with the hazier parts of human instinct. This book will stand up to the intricacies of human misery, the heaviness of history, and the getting through battle for equity and fairness.

It will inspect the human limit with regards to savagery and brutality, as well as our true capacity for sympathy and consideration.

The investigation of the human condition stretches out to the domain of cognizance and the idea of the real world. We will dig into the secrets of discernment, mindfulness, and the philosophical investigations into the idea of presence itself. An excursion provokes us to defy the constraints of our insight and understanding, welcoming us to consider the magical and existential inquiries that have puzzled mankind for a really long time.

At last, the topic of this book is a festival of the human soul, an acknowledgment of the excellence and intricacy of our reality. It is an affirmation of our common weaknesses and qualities, our aggregate yearning for association and significance. The human condition is both a material whereupon we paint the tales of our lives and a mirror in which we see ourselves reflected in the encounters of others.

As we set out on this investigation of the human condition, moving toward it with an open heart and an inquisitive mind is fundamental. It is an encouragement to embrace the vulnerabilities and ambiguities that characterize our reality, to examine the heap viewpoints that have molded how we might interpret the world, and to participate in a discourse that rises above time and culture.

The human condition is a multi-layered and developing subject, one that proceeds to move and challenge us. A subject resounds across ages and across the limits of language and culture. A subject welcomes us to ponder our common process as people, to see the value in the variety of our encounters, and to figure out some mutual interest in our goals and battles.

In the pages that follow, we will experience a different cluster of voices and viewpoints, from old rationalists to contemporary masterminds, from writers and writers to researchers and researchers. These voices offer us looks into the significant and multi-layered nature of the human condition. They help us that the investigation to remember our mankind is a continuous and dynamic undertaking, one that welcomes us to draw in with sympathy, interest, and miracle.

This book is definitely not a conclusive composition on the human condition. It is an assortment of considerations, stories, and bits of knowledge that by and large structure a mosaic of human encounters. A proposing to those try to extend how they might interpret themselves and their kindred individuals. A festival of the common excursion joins all of us, the excursion of investigating the human condition.

As we set out on this excursion, let us recall that the human condition is definitely not a static or fixed substance. It is a no nonsense embroidery that is consistently formed by the narratives we tell, the inquiries we pose, and the associations we manufacture. It is an update that how we might interpret the human condition is a continuous discourse, one that traverses the limits of existence.

In the sections that follow, we will investigate different aspects of the human condition, diving into the profundities of human feelings, the secrets of cognizance, the force of imagination, and the getting through look for importance. We will mull over the significant effect of history and culture on how we might interpret ourselves and the world.

The human condition is a subject that rises above the bounds of any single discipline or point of view. A subject welcomes us to take part in a comprehensive investigation of our reality, one that recognizes the interconnectedness of our physical, close to home, scholarly, and other-worldly aspects.

All through this book, we will experience a different cluster of voices and viewpoints, each offering an interesting focal point through which to see the human condition. These voices come from various periods, societies, and different backgrounds, yet they all offer a typical yearning: to reveal insight into the intricacies of our reality and to rouse reflection and understanding.

The investigation of the human condition is a profoundly private and pensive excursion. It welcomes us to contemplate our own encounters, convictions, and values, and to draw in with the voices of other people who have wrestled with similar crucial inquiries. It is an amazing chance to grow our viewpoints, extend our sympathy, and enhance our viewpoint on the world.

In the sections that follow, we will experience stories and reflections that resound with the full range of human feelings. We will dive into the profundities of adoration and bliss, consider the secrets of dread and distress, and investigate the cryptic landscape of satisfaction and trouble.

The human condition is additionally set apart by our attention to time and mortality. We are aware of the passing idea of our lives, and this mindfulness shapes our decisions, yearnings, and wants. We will ponder the idea of time, the entry of years, and the certainty of progress and misfortune.

1.2 Early use of oil by ancient civilizations.

The usage of oil is a training that traverses centuries, well established throughout the entire existence of human development. The early utilization of oil by old social orders is a demonstration of the genius of our precursors and their capacity to tackle the World's normal assets for many purposes.

One of the earliest recorded uses of oil can be followed back to old Mesopotamia, the support of civilization, where the Sumerians and Babylonians flourished around 3500-2000 BCE. Individuals of this district found the significant properties of bitumen, a thick, tar-like substance that leaked from the beginning. Bitumen was utilized as a cement and waterproofing material, assuming an essential part in the development of old designs like ziggurats and sanctuaries.

The Sumerians and Babylonians likewise utilized bitumen for clinical and remedial purposes. It was used in different treatments and cures, featuring early perceptions of its likely restorative properties. This exhibits the resourcefulness of antiquated civic establishments in perceiving the flexibility of oil-based substances.

Farther toward the east, in the Indus Valley human advancement, one more old society was finding the utility of oil. Archeological proof recommends that individuals of the Indus Valley, who possessed the locale around 2500-1500 BCE (current India and Pakistan), were utilizing sesame oil for cooking. Sesame oil was a wellspring of nourishment as well as assumed a huge part in the culinary practices of the time.

The old Egyptians, a civilization that flourished around 3000-30 BCE, likewise embraced the multi-layered nature of oils. They used different oils, including castor oil, olive oil, and sesame oil, for culinary, restorative, and clinical purposes. Olive oil, specifically, held a respected status in Egyptian culture. It was a dietary staple as well as utilized in strict customs and as a base for fragrances and beauty care products.

The Egyptians were known for their unpredictable excellence ceremonies, and they fostered a large number of corrective items involving oils as key fixings. Oils were utilized to saturate the skin, safeguard it from the brutal desert sun, and upgrade individual preparing. The Ebers Papyrus, an old Egyptian clinical text dating to around 1550 BCE, contains references to the utilization of different oils for restorative purposes, featuring the early acknowledgment of their recuperating properties.

In old India, Ayurveda, an arrangement of customary medication that goes back north of 5,000 years, consolidated oils as fundamental parts in its remedial practices. Ayurvedic texts suggested the utilization of oils, for example, sesame oil, in rubs, known as abhyanga, to advance physical and mental prosperity. These back rubs were accepted to adjust the body's energies and work on generally speaking wellbeing.

The early Greeks likewise embraced the utilization of oils in their day to day routines. Olive oil was a dietary staple as well as utilized for lighting lights and in strict services. In Greek folklore, olive trees were thought of as sacrosanct, and olive oil was related with virtue and godliness. It was utilized as an important product in exchange and assumed a huge part in the way of life of old Greece.

The use of oils stretched out to the Roman Domain, where olive oil was a foundation of both food and culture. The Romans perceived the significance of olive oil in their eating regimen and thought of it as an image of riches and extravagance. It was utilized as a base for different culinary arrangements, as well as a dressing for plates of mixed greens and a fixing for bread.

Olive oil additionally had reasonable applications past the eating table. The Romans involved it as a wellspring of light by consuming it in lights, a training that endured for a really long time. These oil lights enlightened homes, roads, and public spaces, adding to the headway of metropolitan life.

In old China, the utilization of oil was not restricted to culinary and restorative purposes but rather reached out to mechanical developments. The Chinese are credited with imagining the world's most memorable oil well and involving oil as a fuel source. The extraction of oil starting from the earliest stage lighting and warming purposes traces all the way back to the Han Tradition (202 BCE - 220 CE). This early utilization of oil for enlightenment denoted a huge progression in energy innovation.

The authentic proof of oil's initial use by old civic establishments highlights its multi-layered importance in human culture. Oils were necessary to day to day existence, including culinary, therapeutic, strict, and innovative areas. They were wellsprings of food and prosperity as well as images of culture, exchange, and advancement.

The old's comprehension world might interpret oil stretched out past its viable applications. Numerous old societies credited enchanted and otherworldly importance to specific oils. Olive oil, for instance, was related with virtue, mending, and heavenly endowments in different religions and legends. The utilization of oils in strict functions and ceremonies assumed a urgent part in interfacing the material and other-worldly parts of human life.

The early utilization of oils in old civilizations additionally features the significance of exchange and social trade. Oils, particularly olive oil, were important items that were exchanged across districts, cultivating associations and intercultural impacts. The transmission of information and practices connected with oil utilization enhanced the aggregate insight of humankind.

Notwithstanding the useful and social components of oil utilization, the early human's comprehension advancements might interpret oils for restorative intentions was wonderful. These antiquated social orders perceived the recuperating capability of oils, integrating them into different cures and remedial practices. The utilization of oils in back rub, fragrance based treatment, and skin applications for wellbeing and prosperity mirrors an early consciousness of their helpful properties.

While the verifiable record gives experiences into the utilization of oils in antiquated human advancements, it is essential to take note of that the comprehension and usage of oils kept on developing over the long run. The cutting edge time has seen huge headways in oil extraction, refinement, and application, prompting the foundation of businesses and advances that depend on this important asset.

Quite possibly of the most extraordinary period throughout the entire existence of oil use happened during the Modern Transformation, which started in the late eighteenth 100 years.

The disclosure of new techniques for oil extraction, especially the penetrating of oil wells, upset the business. The improvement of the principal business oil well in Pennsylvania, USA, in 1859 by Edwin Drake denoted a defining moment throughout the entire existence of oil investigation and creation.

The Modern Insurgency saw the ascent of the petrol business, with raw petroleum being refined to deliver different items, including lamp oil for lighting, greases for apparatus, and fuel for the incipient auto industry. The boundless accessibility of oil based items changed transportation, industry, and day to day existence, prompting huge cultural changes.

The use of oil as an energy source had significant ramifications for human culture. It filled the fast industrialization and urbanization of the Western world, prompting expanded efficiency and monetary development. The improvement of the gas powered motor in the late nineteenth century upset transportation, preparing for autos, boats, and planes.

The twentieth century saw the development of the petrochemical business, which created a great many items got from petrol, like plastics, engineered elastic, and synthetic substances. These materials reformed assembling, bundling, and purchaser merchandise, changing the manner in which individuals lived and worked.

The international meaning of oil additionally became apparent during the twentieth hundred years. Oil-rich locales, for example, the Center East, acquired worldwide unmistakable quality because of their immense stores of this significant asset. The mission for command over oil assets and the security of oil supply courses assumed a vital part in molding worldwide relations and clashes.

The historical backdrop of oil investigation and extraction is interlaced with mechanical advancements and the improvement of boring strategies. The headway of seismic investigation, seaward penetrating, and upgraded oil recuperation techniques has extended the accessibility of oil holds and expanded creation limit. These progressions have empowered social orders to satisfy developing energy needs and drive financial development.

The cutting edge time has additionally seen the development of sustainable power sources as options in contrast to conventional petroleum derivatives. The quest for practical energy arrangements, for example, sunlight based, wind, and hydropower, mirrors a developing consciousness of the ecological and environment challenges related with the broad utilization of oil and other petroleum derivatives.

The effect of oil on the climate has been a subject of critical concern. The extraction, transportation, and burning of petroleum products have added to air and water contamination, environment interruption, and ozone depleting substance discharges, prompting environmental change. Endeavors to moderate these natural impacts have prompted the advancement of cleaner innovations and environmentally friendly power sources.

All in all, the early utilization of oil by old developments is a demonstration of humankind's cleverness, creativity, and versatility. The acknowledgment of oil's multi-layered utility in culinary, restorative, social, and mechanical spaces established the groundwork for its proceeded with importance in the cutting edge world.

The historical backdrop of oil utilization shows the interconnectedness of social orders, societies, and exchange organizations, as well as the transmission of information and practices across locales and periods. The usage of oils in different structures, from culinary delights to sacrosanct customs, mirrors the rich embroidery of human experience and imagination.

1.3 The first commercial oil well in Pennsylvania.

The principal business oil well in Pennsylvania, known as the Drake Well, holds a huge spot in history as it denoted the introduction of the cutting edge oil industry. Situated close to Titusville, Pennsylvania, this spearheading penetrating activity, started by Edwin L. Drake in 1859, upset how oil was separated and used, prompting significant cultural, financial, and modern changes.

Before the boring of the Drake Well, oil was principally gathered from regular drainages in different districts, including portions of the US, the Center East, and Europe. Native people groups had long realized about these oil leaks and had involved the substance for restorative, grease, and waterproofing purposes. At times, oil leaks were viewed as interests and vacation spots, as they were in many cases joined by petroleum gas vents that could be lighted to deliver a consistent fire.

In any case, the Drake Very much denoted a critical takeoff from these conventional techniques for gathering oil. Edwin L. Drake, a resigned railroad guide, is frequently credited with having the vision and assurance to bore the primary industrially effective oil well. His endeavors to separate oil from the earth upset the business and set up for the oil blast that would follow.

Drake's process started during the 1850s when he was recruited by the Seneca Oil Organization to explore the capability of boring for oil. At that point, oil was fundamentally acquired by digging pits close to oil leaks and gathering the leaking oil. This strategy was wasteful and restricted in its yield, making the cycle exorbitant and impractical.

Drake perceived that penetrating for oil offered a more effective and pragmatic arrangement. He confronted many difficulties, including obtaining the hardware required for boring, convincing financial backers to support the undertaking, and tracking down a reasonable area for penetrating. It was a spearheading adventure that expected imagination and development.

In August 1859, in the wake of confronting various mishaps, Drake and his group effectively struck oil at a profundity of 69.5 feet at the site close to Titusville, Pennsylvania.

The groundbreaking occasion occurred on August 27, and it denoted the introduction of the main financially reasonable oil well ever. The oil delivered from this very much was at first utilized for restorative purposes and as an ointment for hardware.

The outcome of the Drake Very much significantly affected society. It set off a surge of oil investigation and penetrating across the US, prompting the foundation of an industry that would change the worldwide economy. The Drake Very much exhibited the potential for penetrating to get to immense supplies of oil, which were already blocked off through conventional strategies.

One of the key improvements that made the Drake Well's prosperity conceivable was the innovation utilized for penetrating. Drake and his group utilized a boring apparatus that included a steam motor to control the boring tool, making it significantly more proficient than difficult work. This development sped up boring as well as considered penetrating to arrive at a lot more noteworthy profundities, taking advantage of beforehand inaccessible oil holds.

The Drake All around was a quick business achievement. Its underlying creation rate was unobtrusive, yielding around 25 barrels of oil each day, however this was all that could possibly be needed to exhibit the plausibility of business oil penetrating. As insight about the well's prosperity spread, it pulled in the consideration of financial backers and business people anxious to imitate its outcomes.

The progress of the Drake All around prompted the arrangement of various oil organizations, which immediately set off to penetrate their own wells. The region around Titusville, Pennsylvania, turned into a center of oil investigation, with wooden derricks jumping up across the scene as organizations bored for oil. These wooden derricks became notable images of the early oil industry.

The monetary effect of the Drake All around was significant. It prompted a blast in oil creation and a reduction in oil costs, making oil based goods more open to everybody. As oil turned out to be all the more generally accessible, it tracked down new applications in industry, transportation, and lighting.

Perhaps of the main advancement that followed the outcome of the Drake Very much was the inescapable reception of lamp fuel as an illuminant. Preceding the revelation of oil in Pennsylvania, different substances, for example, whale oil and fat oil, were utilized for lighting lights. Lamp fuel, a side-effect of raw petroleum refinement, ended up being a predominant and more reasonable other option. It consumed neatly and gave a consistent and brilliant wellspring of light, going with it a famous decision for indoor and open air brightening.

The accessibility of reasonable lamp oil essentially worked on the personal satisfaction for some individuals, as it made evening exercises and work more achievable and productive. The effect of lamp fuel lighting was especially felt in metropolitan regions, where roads and structures could be enlightened into the evening, upgrading wellbeing and efficiency.

As well as lighting, the oil business likewise assumed an essential part in the improvement of transportation. The utilization of oil based commodities in gas powered motors and as ointments considered the advancement of the car business. The progress from horse-attracted carriages to mechanized vehicles altered transportation and versatility, generally impacting the manner in which individuals lived and worked.

The disclosure of oil in Pennsylvania additionally added to the ascent of significant oil organizations. Organizations like Standard Oil, established by John D. Rockefeller in 1870, merged the oil business and controlled different parts of creation, refining, circulation, and show-casing. This union prompted both expanded productivity and worries about monopolistic practices.

The progress of the Drake Well and the following oil blast had a worldwide effect. The accessibility of reasonable lamp oil and other oil based commodities worked on the personal satisfaction in many regions of the planet. It additionally sped up industrialization and added to the improvement of present day economies.

The effect of the oil business stretched out past lighting and transportation. Oil turned into an imperative asset in the assembling of different items, including plastics, synthetic substances, and manu-factured materials. The adaptability of petrol subsidiaries made them fundamental in endless businesses, from drugs to horticulture.

Nonetheless, the development of the oil business was not without difficulties and contentions. The ecological effect of oil extraction and refinement turned into a subject of worry, as spills and contamination presented dangers to biological systems and human wellbeing. These worries prompted administrative endeavors and expanded examination of the business' practices.

The oil business likewise assumed a huge part in international affairs. Oil-rich areas, especially in the Center East, became central marks of global consideration and clashes. The journey for command over oil assets and the insurance of oil supply courses molded worldwide gov-ernmental issues and security contemplations.

The Drake Well and the resulting oil blast in Pennsylvania were basic crossroads throughout the entire existence of the petrol business. They denoted the start of a groundbreaking period that had an impact on the manner in which individuals lived, worked, and associated with the world. The improvement of the oil business lastingly affected society, financial matters, innovation, and the climate.

The tradition of the Drake Well is as yet apparent today, as it represents the force of advancement, the enterprising soul, and the potential for people to roll out significant improvements on the planet. It fills in as a sign of how a solitary well, penetrated earnestly and creativity, can have expansive results, molding the course of history and impacting the existences of endless individuals.

1.4 Key figures in the early days of the oil industry.

The beginning of the oil business were set apart by an exceptional combination of visionary business visionaries, inventive designers, and smart business pioneers. These key figures assumed significant parts in molding the business, from the revelation of oil to its change into a worldwide financial force to be reckoned with. Their commitments and inheritances made history and keep on impacting the business today.

Edwin L. Drake:

Edwin L. Drake is frequently credited with sending off the cutting edge oil industry with the penetrating of the primary business oil well close to Titusville, Pennsylvania, in 1859. Drake, a resigned railroad guide, was employed by the Seneca Oil Organization to investigate the capability of penetrating for oil. He confronted various difficulties, including the advancement of boring innovation and convincing financial backers to subsidize the endeavor. In spite of these deterrents, Drake effectively struck oil at a profundity of 69.5 feet, denoting a pivotal crossroads throughout the entire existence of the business. His visionary way to deal with penetrating and his assurance to get to oil saves underground set up for the oil blast that would follow.

John D. Rockefeller:

John D. Rockefeller is maybe the most notable figure throughout the entire existence of the oil business. He established Standard Oil in 1870, an organization that would come to overwhelm the business. Rockefeller's business discernment and forceful way to deal with combination made Standard Oil one of the world's biggest and most impressive partnerships. By controlling different parts of creation, refining, dissemination, and showcasing, he accomplished extraordinary productivity and economies of scale. Notwithstanding, Standard Oil's practices additionally prompted worries about monopolistic control, in the long run prompting its separation in 1911 under antitrust regulation. Rockefeller's impact stretched out a long ways past the oil business, and his generous undertakings, including the foundation of the Rockefeller Establishment, left an enduring inheritance.

Ida Tarbell:

Ida Tarbell was a writer and maligner who assumed a critical part in uncovering the acts of the Standard Oil Organization. Her analytical revealing and articles, remarkably the serialized work "The Historical backdrop of the Standard Oil Organization," distributed in McClure's Magazine, shed light on the organization's monopolistic and against cutthroat exercises. Tarbell's careful exploration and convincing composing added to public familiarity with Standard Oil's practices and eventually affected the antitrust endeavors that prompted the organization's separation. Her work is viewed as a crucial crossroads throughout the entire existence of insightful reporting and corporate responsibility.

Colonel Edwin A. Drake:

Colonel Edwin A. Drake, frequently mistook for Edwin L. Drake, was a vital figure in the early improvement of the oil business. While not straightforwardly engaged with penetrating the main business oil well, Colonel Drake assumed a urgent part in propelling boring innovation and the business' initial years. He was a trailblazer in penetrating for salt brackish water, which was utilized to deliver salt through vanishing. All the while, he experienced regular oil leaks and perceived the capability of oil boring. His endeavors and developments added to the developing interest in penetrating for oil in the US.

George Bissell and Jonathan Eveleth:

George Bissell and Jonathan Eveleth were instrumental in the early advancement and improvement of the petrol business. They perceived the capability of oil as a fuel source and got the primary business penetrating lease for oil in Pennsylvania in 1854. Bissell and Eveleth were key figures in upholding for the penetrating of the principal business oil well, which Edwin L. Drake would later attempt. Their vision and support helped establish the groundwork for the oil business' development.

Anthony F. Lucas:

Anthony F. Lucas, an Austrian-conceived American specialist, made a critical commitment to the oil business by penetrating the main significant oil gusher at Spindletop, close to Beaumont, Texas, in 1901. The Spindletop revelation denoted a defining moment in the business, prompting an oil blast in Texas and the Bay Coast. The gusher created a colossal measure of oil, flagging the potential for immense stores in the area and on a very basic level replacing the oil business' topography. Lucas' accomplishment and the ensuing Texas oil blast situated the US as a key part in the worldwide oil market.

Colonel James M. Guffey:

Colonel James M. Guffey was a conspicuous figure in the beginning of the oil business. He assumed a vital part in the turn of events and extension of oil fields in Pennsylvania and Ohio. Guffey was a fruitful free oil maker and purifier, and his business keenness permitted him to flourish in a serious and dynamic industry. His commitments to the advancement of oil assets and the development of free oil organizations in the mid twentieth century were critical.

Howard Hughes Sr.:

Howard Hughes Sr., the dad of the celebrated pilot and business visionary Howard Hughes Jr., was a significant figure throughout the entire existence of the oil business. He established the Hughes Device Organization in 1909, which had some expertise in boring tool innovation and assumed a critical part in propelling penetrating techniques. Hughes Sr's. developments in boring apparatus plan and assembling altogether expanded penetrating proficiency, considering further and more useful wells. His commitments to boring innovation reformed the business and keep on being significant in present day oil investigation.

Mary Ellis O'Hara:

Mary Ellis O'Hara, a spearheading geologist, made critical commitments to the comprehension of oil and gas saves. In the mid twentieth 100 years, O'Hara's geographical skill and hands on work prompted the distinguishing proof of significant oil and gas fields in the US. Her work in planning subsurface designs and concentrating on rock developments established the groundwork for further developed investigation and boring practices. O'Hara's pivotal exploration progressed the study of petrol geography and added to the disclosure of significant oil holds.

George Westinghouse:

George Westinghouse, known essentially for his commitments to the electrical business, likewise assumed a part in the early oil industry. He developed the principal commonsense petroleum gas pipeline, which empowered the effective transportation of flammable gas for lighting and warming. Westinghouse's developments in petroleum gas foundation extraordinarily affected the use of flammable gas, making it an important energy asset.

These vital figures in the beginning of the oil business address a different gathering of people who added to the development and improvement of the business from different points. Their aggregate endeavors, crossing boring innovation, business initiative, news coverage, and development, helped shape the cutting edge oil industry and its persevering through influence on society, innovation, and the worldwide economy. The early trailblazers and pioneers established the groundwork for an industry that proceeds to develop and confront new difficulties in the 21st 100 years.

Chapter 2

The Oil Boom and the Rise of Giants

The historical backdrop of the cutting edge world is inseparably connected to the mission for energy, and maybe no asset has assumed a more critical part than oil. The twentieth century saw an oil blast of uncommon scale, changing economies, international relations, and society itself. This period denoted the ascent of oil goliaths that would come to overwhelm the worldwide energy scene. In this account, we dive into the oil blast and the rise of these titans.

The tale of the oil blast starts in the late nineteenth century when the world was on the cusp of a modern upset. In 1859, Colonel Edwin Drake struck oil in Titusville, Pennsylvania, denoting the introduction of the American oil industry. This occasion put into high gear an outpouring of changes that would reshape the worldwide energy scene. As the interest for oil soar, the oil fields of Pennsylvania turned into the focal point of a prospering industry, drawing large number of examiners, business visionaries, and workers.

The oil rush was not bound to the US. Across the Atlantic, the Russian Realm was likewise seeing a flood in oil creation, eminently in the Baku locale. The quickly industrializing world required this newly discovered energy source to fuel production lines, power trains, and light up homes. Oil was the backbone of progress, and its importance couldn't be exaggerated.

The oil business in its early stages was set apart by an absence of guideline and oversight. The "wildcatters," as they were known, scoured the land for oil, boring wells with little respect for wellbeing or ecological worries. Spills, flames, and mishaps were typical. It was a period of unrestrained energy, corrupt strategic policies, and win and-fail cycles.

Be that as it may, in the midst of the disorder and unrest, a couple of people and organizations stuck out. John D. Rockefeller, the pioneer behind Standard Oil, was one of the most conspicuous figures of the period. His smart business insight permitted him to make an in an upward direction coordinated oil domain that controlled each part of the business, from penetrating to dispersion. By the late nineteenth 100 years, Standard Oil had a close imposing business model on the American oil market, making Rockefeller quite possibly of the most extravagant man ever.

The ascent of Standard Oil and its predominance in the market provoked boundless calls for guideline. The organization's forceful practices and market control prompted the death of the Sherman Antitrust Demonstration in 1890, which meant to diminish monopolistic way of behaving. In 1911, the U.S. High Court requested the separation of Standard Oil into a few more modest organizations, denoting a huge change in the business' scene.

While the US was encountering these extraordinary changes, the remainder of the world was additionally perceiving the capability of oil. In the mid twentieth 100 years, the English Illustrious Naval force progressed from coal to oil as its essential fuel source, a move that expanded the interest for oil around the world. This shift additionally highlighted the essential significance of oil in worldwide undertakings.

The Second Great War carried oil to the front of global legislative issues. The contention was, to some extent, a battle for command over oil-delivering districts, like Mesopotamia (cutting edge Iraq) and the

Caucasus. The English government, perceiving the meaning of these locales, laid out the Iraq Petrol Organization in 1928, tying down admittance to a huge oil hold. Likewise, the Russian Unrest of 1917 saw the nationalization of the Russian oil industry, making ready for the rise of the Soviet Association as a significant oil player.

The interwar period saw the oil business venture into new wildernesses. Organizations like Shell, Texaco, and Bay arose as critical players, working on a worldwide scale. These oil monsters framed collusions, laid out treatment facilities, and created broad appropriation organizations. The Thundering Twenties and the resulting monetary accident of 1929 significantly affected the oil business, with fluctuating oil costs and market instability.

The flare-up of The Second Great War additionally increased the significance of oil. The contention's result, not entirely set in stone by admittance to and control of oil holds. The Hub powers, especially Nazi Germany and Majestic Japan, tried to get oil sources in Eastern Europe and Southeast Asia. The Unified powers, drove by the US, were similarly dedicated to safeguarding their oil advantages. The Clash of Stalingrad, to some extent, spun around control of the Caucasus oil fields, featuring the vital job of oil in wartime.

The consequence of The Second Great War denoted another period for the oil business. The Center East, with its immense oil holds, turned into the focal point of worldwide consideration. The Realm of Saudi Arabia, specifically, held massive potential.

The disclosure of oil in the Bedouin Landmass during the 1930s pulled in American organizations like Chevron and Texaco, and it additionally led to the Middle Eastern American Oil Organization (ARAMCO), a consortium framed by American and Saudi interests.

Soon after the conflict, the interest for oil flooded as post-war reproduction and financial development grabbed hold. The Marshall Plan, which expected to remake Western Europe, and the period of prosperity in the US drove expanded utilization of oil. The worldwide oil market was growing quickly, and with it, the impact of oil monsters.

The 1950s and 1960s were a time of wonderful development and union for the business. Public oil organizations, like Venezuela's PDVSA, Mexico's Pemex, and Saudi Aramco, acquired unmistakable quality. These nations looked to state command over their oil assets and decrease the impact of unfamiliar organizations. OPEC (Association of the Petrol Trading Nations) was laid out in 1960 to oversee oil creation and costs on the whole. OPEC's development denoted a critical change in the worldwide power elements of the oil business.

The 1970s achieved a progression of crucial occasions that would shape the oil scene long into the future. The main oil emergency happened in 1973 when OPEC forced an oil ban on nations that upheld Israel during the Yom Kippur War. This move made oil costs soar and uncovered the world's weakness to oil supply interruptions. It was a reminder for oil-bringing in countries and prompted expanded endeavors to differentiate energy sources and advance energy effectiveness.

The 1979 Iranian Insurgency added one more layer of intricacy to the oil market. The unrest in Iran brought about the fall of the Shah and a sharp decrease in Iranian oil creation. Oil costs flooded once more, and the world encountered a second oil emergency. The occasions of the last part of the 1970s highlighted the international dangers related with oil reliance and prompted endeavors to diminish dependence on Center Eastern oil.

All through the 1980s and 1990s, the oil business kept on advancing. Innovation assumed a vital part in the improvement of profound water penetrating, empowering admittance to beforehand undiscovered stores. Seaward penetrating and the extraction of flighty sources, for example, oil sands and shale oil, opened up new roads for oil investigation. The oil monsters of the twentieth century adjusted to these progressions by enhancing their portfolios and growing their scope.

The finish of the twentieth century saw the resurgence of public oil organizations in oil-delivering nations. Numerous states looked to amplify their command over oil assets and income. Russia, under the administration of President Vladimir Putin, reasserted command over its oil industry, prompting the ascent of state-possessed monsters like Gazprom and Rosneft. Venezuela, under Hugo Chávez, sought after a comparable way with the nationalization of its oil area.

The 21st century brought new difficulties and valuable open doors for the oil business. Environmental change and natural worries became the overwhelming focus, prompting expanded investigation of the oil area's ecological effect. State run administrations and ecological activists pushed for more noteworthy interest in sustainable power and stricter guidelines on discharges. The Paris Understanding of 2015, endorsed by almost 200 nations, focused on restricting an unnatural weather change, flagging a potential shift away from non-renewable energy sources.

The shale upheaval, driven by water powered breaking (deep earth drilling) and flat penetrating, changed the energy scene in the US. This mechanical advancement opened huge stores of oil and petroleum gas in already unavailable shale arrangements. The U.S. turned into a significant oil and gas maker, lessening its reliance on imported products and modifying worldwide energy elements. Oil goliaths like ExxonMobil and Chevron adjusted to this changing scene by putting resources into shale assets and enhancing their tasks.

As the world wrestled with the difficulties of environmental change and the requirement for supportable energy sources, some oil goliaths started to rebrand themselves as energy organizations, recognizing the moving tide. They expanded their interests in sustainable power, electric vehicle framework, and low-carbon advances. BP, for instance, declared its desire to turn into a net-zero fossil fuel byproducts organization by 2050.

The oil business' international importance stays as applicable as could be expected. Oil-rich nations like Saudi Arabia, Russia, and the US keep on employing huge impact on the worldwide stage. International pressures and clashes in the Center East, as well as the Icy locale, feature the essential significance of oil saves. Rivalry for control of energy assets and supply courses stays a vital driver of global relations.

The Coronavirus pandemic in 2020 significantly affected the oil business. Lockdowns and travel limitations prompted a sharp drop in oil interest, making costs fall and making an oversupply emergency. Oil-delivering countries and organizations confronted uncommon difficulties as they wrestled with decreased incomes and the need to adjust to an evolving market. The pandemic featured the business' weakness to outer shocks and the basic of versatility.

2.1 The American oil boom and the rush to Texas.

The historical backdrop of the American oil blast is an account of development, desire, and change. A story unfurled in the late nineteenth and mid twentieth hundreds of years, forming the country's economy as well as its general public and the worldwide energy scene. At the core of this astounding part in history was the Solitary Star State, Texas.

The oil blast in America started with a notable name and a somewhat obscure spot. Colonel Edwin Drake, in 1859, struck oil in Titusville, Pennsylvania, denoting the introduction of the American petrol industry. In any case, it would be Texas that would start to lead the pack in the oil rush, at last acquiring the epithet "The Oil Capital of the World."

In the beginning of the oil business, the US depended vigorously on coal as its essential energy source. Coal fueled processing plants, warmed homes, and drove trains, however its impediments turned out to be progressively apparent. It was awkward to ship, had generally low energy thickness, and delivered bountiful measures of smoke and contamination. The quest for a more proficient energy source prompted the disclosure of oil's huge potential.

Texas had for some time been a place that is known for investigation and opportunity. The state's immense scene was the stuff of legends, with its rambling deserts, moving fields, and vast skylines. Early pilgrims saw Texas as a spot where fortunes could be made, and during the nineteenth 100 years, the fantasy about finding oil there started to flourish.

It was in Corsicana, Texas, in 1894 that the state's most memorable business oil disclosure was made. This disclosure, however humble in scale, put into high gear a chain of occasions that would reshape Texas and the whole country. As fresh insight about this view as spread, it touched off a free for all of investigation and boring across the state, drawing in a rush of fortune searchers and business people. The Corsicana field was only the start, an indication of the abundance that lay secret underneath the surface.

By the turn of the twentieth hundred years, Texas was encountering an oil surge of extraordinary extents. The modest community of Beaumont, situated in Southeast Texas, became ground zero for an epic oil revelation. The Spindletop oil field, found in 1901, denoted the start of another period. This gusher of oil shot many feet up high, flagging the appearance of the American oil blast.

The Spindletop disclosure released a craze of boring and hypothesis. The scene of Beaumont immediately changed from a calm provincial town into a clamoring center point of oil action. Oilmen from everywhere the nation and the world ran to Texas looking for their fortune. The appeal of immense undiscovered stores was overpowering, and the commitment of wealth was sufficient to fuel the fantasies of a large number.

The Spindletop oil blast not just reshaped the fortunes of the people who had a special interest yet additionally brought forth another type of business visionaries. Among them was Pattillo Higgins, a visionary who had long trusted in the oil capability of Texas. He had confronted suspicion and derision however was at last justified by the Spindletop disclosure. Higgins' steadiness paid off, and he turned into a noticeable figure in the Texas oil industry.

The American oil blast was described by an absence of guideline and oversight in its initial days. The "wildcatters," as they were known, scoured the land for oil, penetrating wells with little respect for well-being or natural worries. This period was set apart by spills, flames, and mishaps, mirroring the unrestrained energy, corrupt strategic policies, and win and-fail patterns of the time.

Notwithstanding the turmoil and tumult, a couple of people and organizations caught everyone's eye. John D. Rockefeller, the organizer behind Standard Oil, was one of the most conspicuous figures of the period. His canny business sharpness permitted him to make an in an upward direction coordinated oil realm that controlled each part of the business, from penetrating to conveyance. By the late nineteenth 100 years, Standard Oil had a close imposing business model on the American oil market, making Rockefeller quite possibly of the most extravagant man ever.

The ascent of Standard Oil and its predominance in the market provoked far and wide calls for guideline. The organization's forceful practices and market control prompted the death of the Sherman Anti-trust Demonstration in 1890, which planned to diminish monopolistic way of behaving. In 1911, the U.S. High Court requested the separation of Standard Oil into a few more modest organizations, denoting a huge change in the business' scene.

While the US was encountering these extraordinary changes, the remainder of the world was additionally perceiving the capability of oil. In the mid twentieth hundred years, the English Regal Naval force pro-gressed from coal to oil as its essential fuel source, a move that expanded the interest for oil around the world. This shift additionally highlighted the essential significance of oil in worldwide undertakings.

The Second Great War carried oil to the front of worldwide govern-mental issues. The contention was, to a limited extent, a battle for command over oil-delivering locales, like Mesopotamia (current Iraq) and the Caucasus. The English government, perceiving the meaning of these locales, laid out the Iraq Petrol Organization in 1928, tying down admittance to a tremendous oil save. Likewise, the Russian Upset of 1917 saw the nationalization of the Russian oil industry, preparing for the development of the Soviet Association as a significant oil player.

The interwar period saw the oil business venture into new wilder-nesses. Organizations like Shell, Texaco, and Bay arose as critical players, working on a worldwide scale. These oil goliaths framed coalitions, laid out treatment facilities, and created broad conveyance organizations. The Thundering Twenties and the resulting monetary accident of 1929 significantly affected the oil business, with fluctuating oil costs and market instability.

The flare-up of The Second Great War additionally heightened the significance of oil. The contention's result, not entirely set in stone by admittance to and control of oil holds. The Hub powers, especially Nazi Germany and Magnificent Japan, tried to get oil sources in Eastern Europe and Southeast Asia. The Unified powers, drove by the US, were similarly dedicated to safeguarding their oil advantages. The Clash of Stalingrad, to a limited extent, rotated around control of the Caucasus oil fields, featuring the significant job of oil in wartime.

The fallout of The Second Great War denoted another time for the oil business. The Center East, with its tremendous oil saves, turned into the focal point of worldwide consideration. The Realm of Saudi Arabia, specifically, held colossal potential.

The disclosure of oil in the Middle Eastern Promontory during the 1930s pulled in American organizations like Chevron and Texaco, and it likewise led to the Bedouin American Oil Organization (ARAMCO), a consortium framed by American and Saudi interests.

Soon after the conflict, the interest for oil flooded as post-war reproduction and financial development grabbed hold. The Marshall Plan, which expected to reconstruct Western Europe, and the period of prosperity in the US drove expanded utilization of oil. The worldwide oil market was extending quickly, and with it, the impact of oil goliaths.

The 1950s and 1960s were a time of striking development and solidification for the business. Public oil organizations, like Venezuela's PDVSA, Mexico's Pemex, and Saudi Aramco, acquired unmistakable quality. These nations tried to declare command over their oil assets and lessen the impact of unfamiliar organizations. OPEC (Association of the Oil Sending out Nations) was laid out in 1960 to oversee oil creation and costs aggregately. OPEC's development denoted a huge change in the worldwide power elements of the oil business.

The 1970s achieved a progression of significant occasions that would shape the oil scene for quite a long time into the future. The principal oil emergency happened in 1973 when OPEC forced an oil ban on nations that upheld Israel during the Yom Kippur War. This move made oil costs soar and uncovered the world's weakness to oil supply interruptions. It was a reminder for oil-bringing in countries and prompted expanded endeavors to enhance energy sources and advance energy effectiveness.

The 1979 Iranian Insurgency added one more layer of intricacy to the oil market. The unrest in Iran brought about the fall of the Shah and a sharp decrease in Iranian oil creation. Oil costs flooded once more, and the world encountered a second oil emergency. The occasions of the last part of the 1970s highlighted the international dangers related with oil reliance and prompted endeavors to decrease dependence on Center Eastern oil.

All through the 1980s and 1990s, the oil business kept on developing. Innovation assumed a critical part in the improvement of profound water boring, empowering admittance to beforehand undiscovered stores. Seaward boring and the extraction of capricious sources, for example, oil sands and shale oil, opened up new roads for oil investigation. The oil goliaths of the twentieth century adjusted to these progressions by differentiating their portfolios and extending their range.

The finish of the twentieth century saw the resurgence of public oil organizations in oil-delivering nations. Numerous state run administrations looked to boost their command over oil assets and income. Russia, under the authority of President Vladimir Putin, reasserted command over its oil industry, prompting the ascent of state-possessed monsters like Gazprom and Rosneft. Venezuela, under Hugo Chávez, sought after a comparable way with the nationalization of its oil area.

The 21st century brought new difficulties and open doors for the oil business. Environmental change and ecological worries became the overwhelming focus, prompting expanded investigation of the oil area's natural effect. Legislatures and natural activists pushed for more prominent interest in sustainable power and stricter guidelines on outflows. The Paris Understanding of 2015, endorsed by almost 200 nations, focused on restricting an unnatural weather change, flagging a potential shift away from petroleum products.

2.2 The founding and growth of major oil companies.

The narrative of significant oil organizations is a story of development, venture, and desire. These corporate monsters play had an essential impact in molding the worldwide energy scene, energizing financial development, and impacting international elements. From their humble starting points to their ongoing situations as a portion of the world's biggest and most persuasive enterprises, the excursion of these oil organizations is a convincing story.

The beginning of the oil business were set apart by the endeavors of people and limited scope activities. One of the most famous figures of this time was John D. Rockefeller, whose name would become inseparable from the ascent of significant oil organizations. In 1870, Rockefeller helped to establish Standard Oil of Ohio, which later advanced into the Standard Oil Organization. This was the point at which the American oil industry was in its outset, with early innovation and an unregulated scene.

Rockefeller's splendor lay in his capacity to perceive the potential for combination and reconciliation inside the business. He figured out that controlling all parts of the oil business, from investigation and boring to refining and circulation, could yield critical benefits. Standard Oil quickly extended its range, getting or shaping coalitions with different organizations associated with the oil production network. By 1879, it controlled 90% of the oil refining limit in the US.

The predominance of Standard Oil was not without contention. Its forceful strategic policies, including savage evaluating and monopolistic way of behaving, got under the skin of contenders and the public the same. The organization's command over the oil market prompted calls for antitrust guideline. In 1911, the U.S. High Court requested the separation of Standard Oil, refering to infringement of the Sherman Antitrust Demonstration. This choice denoted a huge defining moment throughout the entire existence of significant oil organizations.

The disintegration of Standard Oil brought about the making of 34 more modest organizations, some of which would proceed to become central parts in the oil business. Names like Exxon (Standard Oil of New Jersey), Chevron (Standard Oil of California), and Mobil (Standard Oil of New York) would become inseparable from the area's development and impact. These recently shaped substances acquired the framework and mastery of their ancestor, empowering them to flourish in a quickly extending market.

The twentieth century saw the rise of significant oil organizations past the US, and their development would be portrayed by worldwide extension and broadening. English Petrol (BP), established in 1909, was one such organization that would turn into an easily recognized name. Initially settled as the Old English Persian Oil Organization, it tied down privileges to investigate for oil in Iran, which was wealthy in oil saves.

The Second Great War assumed a critical part in the ascent of significant oil organizations. The contention, driven by the requirement for fuel and greases, increased the interest for oil. Subsequently, organizations like Exxon and Regal Dutch Shell saw their fortunes rise. The requirement for dependable admittance to oil assets prompted a reconfiguration of worldwide influence elements, with nations and organizations competing for control of oil-rich districts.

The interwar period saw the oil business' venture into new boondocks. Significant oil organizations were enhancing their activities, putting resources into investigation and improvement in nations like Venezuela, Mexico, and Indonesia. The revelation of oil in these districts opened up new roads for development and expanded the impact of significant oil organizations on a worldwide scale.

The mid-twentieth century got a shift the overall influence inside the business. Significant oil organizations from the US started to confront contest from worldwide players, especially from the Center East. The making of the Association of the Petrol Trading Nations (OPEC) in 1960 denoted an aggregate exertion by oil-creating countries to apply more prominent command over oil creation and evaluating. This move tested the predominance of significant oil organizations and brought about a shift of force towards oil-sending out nations.

In spite of this test, significant oil organizations adjusted and extended their range. The worldwide interest for oil kept on developing, driven by financial turn of events, urbanization, and the ascent of the auto business. Organizations like Exxon and Shell made huge interests in innovative work to further develop boring advances, improve extraction processes, and extend their stores.

During the 1970s, significant oil organizations confronted a progression of difficulties that would test their strength and flexibility. The oil emergencies of 1973 and 1979, set off by political contentions and supply disturbances in the Center East, prompted taking off oil costs and supply vulnerabilities. These emergencies significantly affected significant oil organizations and the worldwide economy, inciting endeavors to lessen reliance on Center Eastern oil.

Significant oil organizations started to expand their energy portfolios because of developing natural worries and calls for more prominent supportability. They extended their interests into petroleum gas, petrochemicals, and sustainable power sources. Organizations like BP rebranded themselves as "past oil," flagging their obligation to cleaner and more manageable energy arrangements.

The late twentieth century likewise denoted a resurgence of public oil organizations in oil-delivering nations. States tried to amplify their command over oil assets and income. Russia, under President Vladimir Putin's authority, reasserted command over its oil industry, prompting the ascent of state-claimed monsters like Gazprom and Rosneft. Venezuela, under Hugo Chávez, sought after a comparative way with the nationalization of its oil area.

The 21st century introduced another time for significant oil organizations, described by expanding worries about environmental change and the requirement for a progress to low-carbon energy sources. These organizations confronted mounting strain to decrease their fossil fuel byproducts, put resources into environmentally friendly power, and embrace more economical practices. The Paris Understanding of 2015, endorsed by almost 200 nations, focused on restricting an unnatural weather change, flagging a potential shift away from non-renewable energy sources.

Significant oil organizations, like ExxonMobil and Chevron, adjusted to the changing scene by putting resources into shale oil and gas assets, which changed the energy elements of the US. The shale transformation, driven by water powered cracking (deep oil drilling) and level boring, opened immense stores of oil and flammable gas in beforehand unavailable developments. The US turned into a significant oil and gas maker, lessening its reliance on imported products.

As the world wrestled with the difficulties of environmental change and the requirement for feasible energy sources, significant oil organizations progressively situated themselves as energy organizations. They expanded their interests in environmentally friendly power, electric vehicle foundation, and low-carbon advancements. Organizations like TotalEnergies, for instance, have focused on progressing towards a more different energy portfolio, perceiving the requirement for cleaner and more practical arrangements.

The Coronavirus pandemic in 2020 significantly affected significant oil organizations. Lockdowns and travel limitations prompted a sharp drop in oil interest, making costs plunge and making an oversupply emergency. Significant oil organizations were confronted with the double test of diminished incomes and the basic to adjust to an evolving market. The pandemic highlighted the business' weakness to outer shocks and the need of versatility.

Looking forward, the fate of significant oil organizations stays questionable. The progress to a low-carbon economy is picking up speed, with legislatures and organizations resolving to diminish outflows and put resources into cleaner energy choices. The push for electric vehicles, environmentally friendly power, and feasible practices keeps on testing the strength of significant oil organizations.

In any case, significant oil organizations will probably stay fundamental to the worldwide energy scene for years to come. Their huge framework, specialized skill, and experience make them crucial players in giving energy answers for meet the world's developing necessities.

These organizations are adjusting to the changing climate by putting resources into cleaner advancements, carbon catch, and economical works on, lining up with developing cultural and natural assumptions.

The establishing and development of significant oil organizations address a convincing story of advancement, combination, and versatility. From the beginning of wildcatters and trailblazers to the development of global partnerships, these organizations play had a focal impact in forming the world's energy scene. Their development and reaction to changing worldwide elements highlight the always present requirement for ground breaking techniques and supportable practices in the mission for energy arrangements in the 21st hundred years and then some.

2.3 The development of oil infrastructure and pipelines.

The improvement of oil framework and pipelines has been a basic part of the worldwide energy industry, working with the extraction, transportation, and conveyance of one of the world's most crucial assets. Throughout the long term, the development of this framework plays had a urgent impact in molding economies, affecting international relations, and fulfilling the steadily developing need for oil.

The beginning of the oil business were set apart by an absence of foundation and innovation. During the nineteenth 100 years, oil was basically separated through straightforward, shallow wells, and shipped in barrels by horse-drawn carts. This simple methodology restricted the size of oil creation and dissemination, as well as the distance over which it very well may be moved.

Notwithstanding, as the business developed and oil holds were found in new districts, the requirement for a more productive and sweeping framework became obvious. One of the key advancements was the presentation of pipelines. The principal known oil pipeline was developed in the US in 1865 in Pennsylvania, intended to move oil brief good ways from wells to capacity tanks or processing plants. This undeniable the start of a huge change in the manner oil was moved.

Pipelines immediately became essential to the business, empowering the transportation of raw petroleum over longer distances with more prominent effectiveness and wellbeing. The pipeline networks extended as more oil fields were found and taken advantage of, interfacing creation regions with refining focuses and circulation center points. This development in foundation considered the proficient progression of oil to fulfill the rising worldwide need.

In the US, the improvement of pipeline framework assumed a significant part in the development of the oil business. The development of the Trans-The Frozen North Pipeline Framework (TAPS) during the 1970s was a fantastic designing accomplishment. Crossing 800 miles from Prudhoe Straight to Valdez, Gold country, TAPS was intended to ship unrefined petroleum from the state's North Slant to sans ice waters for send out. This task opened up immense oil holds as well as addressed a wonderful accomplishment in pipeline designing.

In the worldwide field, the oil business saw the development of key transportation courses and pipelines that associated oil-creating districts with worldwide business sectors. The Waterway of Hormuz, a limited stream among Iran and the Bedouin Landmass, turned into a basic way for oil shipments from the Persian Bay. The development of monstrous big hauler armadas and stacking terminals at ports in the locale considered the transportation of oil to objections all over the planet.

Another significant pipeline project was the Trans-Siberian Pipeline, which associated the oil fields of Western Siberia with European business sectors. Extending large number of miles, this pipeline was a basic channel for Russian oil commodities to Europe, featuring the international meaning of oil framework.

In the Center East, the development of pipelines was instrumental in associating oil-creating nations with the Mediterranean Ocean, the Red Ocean, and the Middle Eastern Ocean, empowering the commodity of oil to worldwide business sectors. The essential significance of these pipelines and their insurance turned into a significant focal point of worldwide legislative issues and security.

The development of oil framework and pipelines has not been without difficulties and dangers. Natural worries, for example, oil slicks and breaks, have brought up issues about the security and supportability of these frameworks. Significant oil slicks, similar to the Exxon Valdez in 1989 and the Deepwater Skyline in 2010, highlighted the overwhelming effect that foundation disappointments can have on biological systems and networks.

Security concerns likewise arose as pipelines became focuses for damage and psychological warfare in areas of political precariousness. Struggle inclined regions, like Nigeria's Niger Delta and the Center East, saw assaults on pipelines and other oil foundation, disturbing the progression of oil and compromising worldwide energy security.

Endeavors to alleviate these difficulties have prompted headways in pipeline innovation and wellbeing conventions. Checking frameworks, early admonition discovery, and further developed materials have been carried out to decrease the gamble of spills and holes. Moreover, global associations and state run administrations have done whatever it takes to lay out guidelines and principles for the development and activity of pipelines to guarantee wellbeing and ecological security.

The worldwide interest for oil proceeds to rise, and with it, the requirement for additional development of oil framework and pipelines. The Icy district, with its true capacity for tremendous oil saves, has turned into a point of convergence for future turn of events. Nonetheless, the brutal environment and natural responsive qualities present special difficulties for building and keeping up with pipelines in this remote and biologically sensitive region.

The push for cleaner and more feasible energy sources has likewise prompted the advancement of pipelines for the transportation of petroleum gas. Petroleum gas pipelines have turned into a fundamental part of the energy blend, shipping this cleaner-consuming non-renewable energy source to control plants, ventures, and homes.

The improvement of oil framework and pipelines has not been restricted to the transportation of unrefined petroleum alone. Pipelines have likewise been critical in the appropriation of refined oil based commodities, like gas, diesel, and fly fuel. These pipelines guarantee that the items are productively moved from treatment facilities to dissemination focuses, giving a consistent stock to meet purchaser and modern requirements.

The worldwide interconnectedness of energy markets has made pipelines a fundamental part of the worldwide economy. They have worked with the progression of oil from distant extraction locales to shoppers all over the planet. Pipelines have been an essential figure the improvement of significant oil organizations and the extension of their tasks. The capacity to move oil effectively has permitted these organizations to get to different wellsprings of unrefined petroleum and extend their venture into worldwide business sectors.

The international ramifications of oil foundation and pipelines couldn't possibly be more significant. Admittance to, and control of, energy assets is a critical consider worldwide relations and clashes. Significant powers and countries have tried to tie down pipelines and key courses to guarantee the dependable progression of oil. The Waterway of Hormuz, the Druzhba pipeline in Europe, and the pipelines going through Ukraine are instances of locales and courses that have been vital to international moving and pressures.

As the world pushes toward a more feasible energy future, the job of oil foundation and pipelines might develop. The change to cleaner energy sources, like sustainable power and electric vehicles, could diminish the interest for oil, affecting the requirement for broad pipeline organizations. Organizations engaged with the oil and gas industry might have to adjust their methodologies and framework to line up with these evolving elements.

The improvement of oil framework and pipelines is a demonstration of human inventiveness and designing ability. These multifaceted organizations of pipelines have associated remote oil fields to business sectors across landmasses, molding economies, international affairs, and the worldwide energy scene. Their development and flexibility have been integral to meeting the world's energy needs and guaranteeing the progression of this essential asset. As the energy scene keeps on developing, the fate of oil foundation and pipelines stays a basic region of the planet energy progress and the mission for economical energy arrangements.

2.4 The impact of oil on economic expansion.

The revelation and use of oil significantly affect financial extension around the world, molding the development and improvement of countries, enterprises, and individual vocations. Oil, frequently alluded to as "dark gold," isn't just an indispensable wellspring of energy yet additionally a vital driver of financial development, industrialization, and innovative advancement. This paper investigates the multi-layered effect of oil on monetary extension and its broad outcomes.

The excursion starts during the nineteenth century when the primary business oil all around was penetrated by Colonel Edwin Drake in Titusville, Pennsylvania, in 1859. This occasion denoted the origin of the advanced oil industry and established the groundwork for a financial change that would reshape the world. At first, oil was removed basically for use in lamp fuel lights, giving a more brilliant and more proficient wellspring of lighting contrasted with conventional options like whale oil.

As the nineteenth century gave way to the twentieth, the interest for oil developed dramatically. Industrialization and the extension of railways, transportation, and assembling drove the requirement for oil items like ointments and fuel for apparatus. Oil's flexibility and energy thickness made it a key empowering influence of these turns of events. The presentation of the gas powered motor in the late nineteenth century made ready for the large scale manufacturing of autos and made a voracious craving for fuel.

In the US, the oil business went through quick development. Significant oil revelations, like the Spindletop gusher in Texas in 1901, filled a blast that pulled in business people, examiners, and financial backers. Towns and urban communities jumped up around newfound oil fields, prompting expanded business open doors and monetary development. The change of spots like Beaumont, Texas, into clamoring oil center points exhibited the significant effect of the oil business on neighborhood economies.

The development of the American oil industry likewise made fortunes for people and families, for example, the Rockefellers, who created massive financial wellbeing through the establishing of Standard Oil. John D. Rockefeller's business intuition and vision for union permitted Standard Oil to control most of the nation's oil refining limit by the late nineteenth 100 years. While this predominance prompted antitrust guideline and the possible separation of the organization, it highlighted the monetary power and impact of the oil business.

In the mid twentieth 100 years, the effect of oil reached out past the US to the worldwide stage. The coming of the vehicle and the improvement of the flying business enhanced the interest for oil on a worldwide scale. This request affected different areas of the economy, including assembling, transportation, and framework advancement.

The development of the car business, especially in the US, prodded financial extension through the foundation of immense creation offices, the formation of millions of occupations, and the improvement of many auxiliary organizations. As the working class extended and vehicle possession turned out to be more available, the interest for cars and their fuel, gas, added to vigorous financial development and success.

Notwithstanding the auto business, flight assumed a pivotal part in driving financial development. The development of the plane by the Wright siblings in 1903 opened up new open doors for movement, business, and exchange. Aeronautics' reliance on oil-based flight energizes, like flying gas and later stream fuel, laid out the significance of the oil business in molding the cutting edge world. Air transportation worked with the development of products, individuals, and thoughts, fortifying worldwide associations and economies.

The industrialization of horticulture likewise depended on the energy got from oil. Work vehicles, controlled by gas powered motors and energized by gas or diesel, upset cultivating rehearses. These machines expanded farming efficiency and assumed a significant part in taking care of developing populaces. Subsequently, the agribusiness area had the option to deliver surplus food, adding to financial development and food security.

The extension of the oil business was not restricted to the US. Worldwide players like Regal Dutch Shell, English Oil (BP), and Exxon-Mobil developed into worldwide goliaths. These organizations laid out activities in oil-rich locales across the globe, framing associations and partnerships that drawn out their span and impact.

The international effect of oil on financial extension can't be put into words. The twentieth century saw the essential significance of oil-delivering areas, particularly the Center East. The Second Great War and The Second Great War were, to some degree, battles for command over oil saves. The fights in the Center East, especially for admittance to oil assets in Mesopotamia (current Iraq) and the Caucasus, uncovered the basic job of oil in worldwide relations and clashes.

Command over oil saves gave countries huge influence on the worldwide stage, permitting them to shape political unions and apply impact over energy-subordinate countries. The foundation of the Association of the Oil Sending out Nations (OPEC) in 1960 addressed an aggregate exertion by oil-delivering countries to declare command over oil creation and evaluating, a move that tested the predominance of significant oil organizations and reshaped the worldwide energy scene.

The 1970s were set apart by two significant oil emergencies that had extensive ramifications for monetary development. The primary emergency happened in 1973 when OPEC forced an oil ban on nations that upheld Israel during the Yom Kippur War.

This move made oil costs soar, prompting monetary disturbance in oil-bringing in countries. The emergency highlighted the world's weakness to oil supply interruptions and provoked endeavors to lessen reliance on Center Eastern oil.

The second oil emergency occurred in 1979 because of the Iranian Transformation, which brought about the fall of the Shah and a sharp decrease in Iranian oil creation. The subsequent stockpile disturbances and taking off oil costs significantly affected the worldwide economy, adding to stagflation and monetary unsteadiness.

The oil emergencies of the 1970s prompted a crucial change in financial reasoning and strategy. Policymakers perceived the need to improve energy security and lessen dependence on oil from politically unsound areas. This acknowledgment brought about drives pointed toward further developing energy proficiency, putting resources into elective energy sources, and broadening the energy blend.

The improvement of oil foundation and pipelines assumed a basic part in supporting the oil business' development and guaranteeing the effective dispersion of oil and its items. Pipeline networks extended to interface oil-creating districts with treatment facilities, circulation center points, and global business sectors. These pipelines worked with the transportation of raw petroleum, refined items, and flammable gas, empowering the business to satisfy the rising worldwide need for energy.

The development of the Trans-The Frozen North Pipeline Framework (TAPS) during the 1970s addressed a stupendous designing accomplishment. Extending 800 miles from Prudhoe Sound to Valdez, Gold country, TAPS was intended to move unrefined petroleum from the state's North Incline to sans ice waters for trade. This task opened up tremendous oil holds as well as exhibited the capacity of current designing to conquer testing natural circumstances.

In the worldwide field, key pipelines and transportation courses associated oil-creating districts with worldwide business sectors. The Waterway of Hormuz, a tight stream among Iran and the Middle Eastern Landmass, turned into a basic path for oil shipments from the Persian Bay. The development of monstrous big hauler armadas and stacking terminals at ports in the area considered the transportation of oil to objections all over the planet.

Pipelines have additionally been instrumental in the dispersion of refined oil based goods, like gas, diesel, and fly fuel. These pipelines guarantee the proficient vehicle of items from treatment facilities to dispersion focuses, giving a consistent stockpile to meet shopper and modern requirements.

The worldwide interconnectedness of energy markets has made pipelines a fundamental piece of the worldwide economy. They have worked with the progression of oil from far off extraction destinations to customers all over the planet.

This interconnectedness has empowered the advancement of significant oil organizations, permitting them to get to assorted wellsprings of unrefined petroleum and grow their venture into global business sectors.

The effect of oil on financial development has not been without difficulties and dangers. Natural worries, for example, oil slicks and holes, have brought up issues about the wellbeing and maintainability of oil foundation and pipelines. Significant oil slicks, similar to the Exxon Valdez in 1989 and the Deepwater Skyline in 2010, highlighted the overwhelming effect that foundation disappointments can have on biological systems and networks.

Security concerns additionally arose as pipelines became focuses for damage and psychological warfare in areas of political shakiness. Struggle inclined regions, like Nigeria's Niger Delta and the Center East, saw assaults on pipelines and other oil foundation, disturbing the progression of oil and compromising worldwide energy security.

Endeavors to alleviate these difficulties have prompted progressions in pipeline innovation and security conventions. Observing frameworks, early admonition recognition, and further developed materials have been carried out to decrease the gamble of spills and breaks. Moreover, global associations and states have done whatever it may take to lay out guidelines and norms for the development and activity of pipelines to guarantee wellbeing and ecological insurance.

Chapter 3

Fuelling the Transportation Revolution

The historical backdrop of human civilization is firmly interlaced with the advancement of transportation. From the earliest long stretches of human life, the need to move individuals and merchandise has driven advancement and cultural advancement. While different methods of transportation, like strolling, creature drawn vehicles, and cruising ships, play played basic parts since forever ago, the transportation unrest that unfurled with the approach of the gas powered motor, filled by oil, remains as perhaps of the most groundbreaking and persevering through section in the tale of human versatility.

This paper digs into the entrancing story of how oil, as a fuel source, lighted the transportation upheaval, investigating its significant effect on the manner in which individuals and merchandise move, its commitments to financial turn of events and globalization, and the difficulties and open doors it presents in a period of expanding natural mindfulness and economical transportation choices.

The Gas powered Motor and the Introduction of Present day Transportation

The gas powered motor (ICE), an innovation that controlled the transportation transformation, arose as a unique advantage in the late nineteenth hundred years. It denoted a takeoff from conventional types of motion, for example, steam motors and muscle power. The vital element of the ICE was its capacity to change over the energy contained in fluid fuel, essentially gas and diesel, into mechanical movement. This development opened another period of transportation portrayed by adaptability, speed, and flexibility.

The far and wide reception of the ICE prompted the advancement of different vehicles, going from cars to cruisers, trucks, and transports. In the mid twentieth 100 years, trailblazers like Henry Passage and Karl Benz spearheaded the large scale manufacturing of cars, making individual portability open to an expansive section of the populace. The reasonableness and accommodation of vehicles upset individual transportation, empowering people to travel autonomously, access open positions, and investigate new skylines.

The ascent of the vehicle business changed urban communities and metropolitan preparation. The requirement for foundation, like streets, interstates, and leaving offices, arose to oblige the developing number of vehicles. The suburbanization of networks, empowered by the simplicity of driving, reshaped the social and financial structure holding the system together.

Simultaneously, the shipping business experienced critical development. Trucks turned into the foundation of business transportation, conveying products productively and interfacing locales and markets. Shipping organizations jumped up across the US, changing planned operations and supply chains. The shipping business' development was intently attached to the accessibility and moderateness of diesel fuel, which offered high energy thickness and productivity.

The railroad business, as well, tackled the force of the gas powered motor. Diesel trains, which supplanted steam motors, offered better proficiency, diminished working expenses, and expanded dependability. This progress empowered railways to give financially savvy transportation benefits and cultivated monetary advancement by working with the development of merchandise over significant distances.

The Ascent of Aeronautics and Its Reliance on Oil

The transportation unrest reached out to the skies with the advancement of the aeronautics business. Controlled flight, drove by pioneers like the Wright siblings, opened up new boondocks and changed the manner in which individuals navigated landmasses and seas. Aeronautics depended intensely on the gas powered motor, explicitly the flight fuel motor.

The requirement for superior execution flight gas drove propels in refining innovation. The presentation of leaded flight gas took into account expanded motor pressure proportions and power yield. This improvement assumed a significant part in the outcome of early airplane and the development of the flying business.

The Second Great War and The Second Great War sped up the improvement of aeronautics innovation. The conflicts filled in as proving grounds for airplane and impetus frameworks, prompting advancements like turbojet motors. After the conflicts, the flight business moved its concentration to business air travel, with carriers like Container American World Aviation routes (Skillet Am) and TWA driving the way. The development of business flight associated individuals, societies, and economies on a worldwide scale, transforming air travel into an image of innovation and progress.

Fly flight, which turned into a reality during the 1950s, denoted one more achievement in the transportation upset. Stream motors, controlled by lamp oil based flying fuel, upset air travel with their speed, productivity, and reach. Stream carriers like the Boeing 707 and the Douglas DC-8 united the world, making global travel more available and reasonable. The far reaching reception of fly airplane laid out the establishment for the cutting edge worldwide economy, empowering the quick development of individuals and products across mainlands.

Delivery and Oil: An Oceanic Organization

The transportation unrest likewise stretched out to the ocean, changing sea business and worldwide exchange. The oceanic business has been a foundation of globalization, and the reception of oil as a fuel source assumed a crucial part in making it conceivable. The change from coal to oil-controlled ships further developed effectiveness and considered longer journeys without the requirement for successive refueling.

The SS Savannah, a crossover cruising steamship fueled by coal and wood, is in many cases considered one of the earliest vessels to utilize steam impetus. Notwithstanding, it was the Imperial Naval force's choice to embrace oil as a fuel hotspot for warships in the late nineteenth century flagged the start of the oceanic shift toward oil. Oil's benefits over coal included higher energy thickness, more straightforward stockpiling, and quicker refueling, pursuing it an optimal decision for maritime vessels.

Dealer delivering went with the same pattern, taking on oil-terminated motors that gave more noteworthy reach and adaptability. The change to oil-controlled ships harmonized with the development of the oil business, guaranteeing a dependable stock of marine fuel. The improvement of huge big haulers fit for shipping oil across seas additionally worked with the worldwide development of oil based goods and unrefined petroleum.

The development of oil big hauler armadas carried new efficiencies to oceanic transportation. The capacity to ship enormous amounts of oil made it financially savvy to move oil from creation locales to far off business sectors. This improvement not just assumed a critical part in oil exchange yet additionally significantly affected worldwide trade, as it considered the productive development of merchandise and assets all over the planet.

The Panama Waterway, a basic oceanic entryway, went through huge redesigns and extensions in the mid twentieth 100 years to oblige bigger vessels, including oil big haulers. The channel's significance in working with worldwide exchange was highlighted by its part in associating the Atlantic and Pacific Seas, lessening travel times and transportation costs.

The waterway's modernization further sped up the globalization of the oceanic business and the development.

3.1 The advent of the automobile and its reliance on oil.

The creation of the car remains as perhaps of the most extraordinary crossroads throughout the entire existence of transportation and present day culture. The ascent of the vehicle not just reshaped the manner in which individuals moved and associated yet in addition significantly affected economies, enterprises, and metropolitan preparation. This paper digs into the coming of the car and its crucial job in upsetting individual versatility, with a specific spotlight on its dependence on oil as a fuel source and the sweeping results this relationship had on the world.

The Creation of the Auto

The improvement of the vehicle was a summit of a few mechanical developments and the visionary work of innovators and specialists. While the idea of a self-pushed vehicle had been investigated for quite a long time, it was during the late nineteenth and mid twentieth hundreds of years that the car as far as we might be concerned today started to come to fruition.

The primary self-moved street vehicle is frequently credited to Nicolas-Joseph Cugnot, a French specialist, who constructed a steam-fueled tricycle in 1769. In any case, this early development was to a greater degree a model rather than a useful vehicle. It was only after the late nineteenth century that huge headway was made in the advancement of the vehicle.

Karl Benz, a German designer, is broadly perceived as the creator of the cutting edge gas controlled car. In 1885, he fabricated the Benz Patent-Motorwagen, a three-wheeled vehicle controlled by a gas powered motor. This development is viewed as the main genuine car since it included a lightweight, gas powered motor and was intended for individual transportation.

The Benz Patent-Motorwagen was trailed by other spearheading vehicles, for example, the Duryea Engine Cart, worked by siblings

Charles and Plain Duryea in the US in 1893. This was the principal American-fabricated car, controlled by a gas motor. These early cars were regularly hand tailored and costly, restricting their openness to a little, rich portion of the populace.

The Job of Henry Passage and the Large scale manufacturing Insurgency

The far and wide reception of the car required advancements in both innovation and assembling. The urgent crossroads in the car's set of experiences accompanied the presentation of the sequential construction system by Henry Portage. Passage's development changed how cars were created, making them more reasonable and open to the overall population.

In 1908, Passage presented the Model T, a historic vehicle that highlighted a straightforward plan, unwavering quality, and moderateness. The Model T, frequently alluded to as the "Dilapidated car," turned into an image of the large scale manufacturing upset. Passage's utilization of mechanical production system strategies, like compatible parts and division of work, essentially diminished the expense of assembling, permitting the Model T to be delivered in huge amounts.

The presentation of the Model T denoted the start of individual versatility for the general population. It turned into the main genuinely reasonable car, with a value that consistently diminished throughout the long term. Portage's vision was to make the car available to the typical American, and he prevailed with regards to doing as such. The moderateness and common sense of the Model T made it a colossal achievement, with millions sold over its creation run.

The vehicle's effect on American culture and economy was monstrous. It achieved a change in perspective in private transportation and modified the manner in which individuals lived and worked. The expanded portability given by the car changed metropolitan and provincial regions, as individuals could now live farther from their working environments and drive all the more without any problem. This, thus, prodded suburbanization and the advancement of framework like streets, interstates, and service stations.

The Auto and Monetary Development

The development of the auto business assumed an essential part in monetary extension. The assembling and deals of cars made great many positions, straightforwardly and by implication. The interest for autos invigorated the creation of steel, glass, elastic, and different parts, prompting the development of related businesses.

The auto business' extension had far reaching influences across the economy. It prodded interest in innovative work, prompting advancements in designing, plan, and materials. The improvement of the oil business to give gas and different fills additionally added to the monetary effect of the auto.

The car likewise changed the retail area. Vehicle sales centers turned into an omnipresent component of metropolitan and rural scenes, offering an extensive variety of car models and supporting choices. Auto related organizations, from fix shops to parts providers, multiplied, adding to the financial biological system.

Moreover, the car business' requirement for a huge organization of streets and expressways prompted framework improvement. Legislatures at different levels put resources into the development and support of transportation organizations to oblige the developing number of vehicles. This interest in foundation further developed transportation as well as made positions and animated neighborhood economies.

The effect of the vehicle stretched out past the US, with different nations additionally encountering monetary development and industrialization. In Europe, vehicle fabricating focuses arose in nations like Germany, France, and the Unified Realm. Japan's auto industry, at first impacted by American models, quickly formed into a significant worldwide player.

The auto business added to a blast in the assembling area. The requirement for steel, glass, elastic, and different materials to fabricate vehicles prompted the development of these ventures. Large scale manufacturing strategies created for vehicles, for example, the sequential construction system, were subsequently embraced by different businesses, further advancing monetary development and advancement.

The Moderateness of the Auto

One of the key elements driving the auto's prosperity and financial effect was its rising reasonableness. Henry Portage's Model T, which was evaluated at $825 in 1908, went through a progression of cost decreases because of advances in assembling procedures, making it more open to a more extensive crowd. By 1924, the cost of a Model T had dropped to $260, and it turned into the vehicle of decision for innumerable Americans.

The moderateness of the car gained it an image of headway and a method for up portability. Working class families could now try to claim a vehicle, which conceded them the opportunity to travel, investigate new spots, and visit loved ones. The versatility given by the car likewise set out new relaxation open doors, as individuals could take get-aways, visit public stops, and go to social and games.

The inescapable reception of the car likewise significantly affected provincial networks. Ranchers and country inhabitants accessed a more extensive scope of administrations and products, as they could now go to local towns and urban communities all the more without any problem. Rustic streets were improved to oblige auto traffic, further interfacing country regions to the more extensive economy.

The vehicle's reasonableness and openness likewise energized the development of the travel industry. Individuals could now leave on travels, investigating their own nations and in any event, wandering abroad. The "Incomparable American Excursion" turned into an esteemed practice, with notable parkways like Highway 66 and the Pacific Coast Interstate becoming inseparable from experience and investigation.

The Financial Effect of Oil in the Auto Business

The connection between the vehicle and the oil business is a fundamental part of the transportation upheaval. Oil, as gas and later diesel fuel, turned into the essential wellspring of energy for the gas powered motors driving cars.

Gas immediately turned into the fuel of decision for vehicles. It offered a high energy thickness, permitting vehicles to travel significant distances without successive refueling. Fuel was likewise somewhat simple to store and ship, going with it a down to earth decision for individual vehicles.

The developing interest for gas prompted huge development in the oil business. Petroleum treatment facilities jumped up to deal with raw petroleum into fuel and other oil based commodities. The improvement of pipelines and transportation networks considered the effective circulation of gas to energizing stations the nation over.

The collaboration between the auto and the oil business was instrumental in the development and enhancement of the two areas. As additional individuals embraced cars, the interest for gas expanded, which, thus, prodded the development of the oil business to fulfill this need. The association between the two enterprises added to their financial success.

The accessibility of fuel and the simplicity of refueling at service stations made the auto a down to earth and helpful method of transportation. Service stations became normal apparatuses in metropolitan and rural regions, offering a promptly open wellspring of fuel for drivers. This framework upheld the development of the vehicle business and made extremely long travel more attainable.

The Job of Diesel in Business Transportation

While gas controlled individual cars, diesel fuel assumed an essential part in the extension of business transportation. Diesel motors, known for their proficiency and force, were appropriate for fueling weighty trucks and transports. The reception of diesel motors for business vehicles changed coordinated operations, empowering the productive development of merchandise.

The shipping business turned into a basic part of the transportation upset. Trucks filled in as the foundation of business transportation, conveying products from makers to retailers and customers. The extension of the shipping business added to the development of the assembling and retail areas, driving financial development.

3.2 The role of oil in the expansion of the aviation industry.

The flying business has a celebrated history, set apart by development, investigation, and mechanical headways. One of the critical drivers of this industry's prosperity has been its dependence on oil as an essential wellspring of energy. From the earliest propeller-driven airplane to the present powerful fly carriers, oil plays had a basic impact in controlling the motors that empower air travel. This paper dives into the vital job of oil in the extension of the flying business, investigating how this association has changed the manner in which individuals and merchandise move, driven monetary development, and molded worldwide network.

Beginning of Flight: The Change from Propellers to Stream Motors

The beginning of avionics saw the utilization of lightweight, aircooled motors principally filled by gas. These motors fueled spearheading airplane like the Wright siblings' Flyer and the Curtiss Jenny biplanes. Fuel, with its high energy thickness and instability, was appropriate for these early airplanes, which depended on the gas powered motor to produce push.

As avionics innovation progressed, the business started to investigate all the more remarkable and proficient motors to accomplish higher velocities, heights, and payloads. During the interwar period, flying producers created airplane with bigger, fluid cooled motors, frequently powered by flight gas, a particular high power fuel intended for airplane motors.

The progress from cylinder motors to stream drive addressed a stupendous change in flight. The advancement of the turbojet motor by Plain Shave in the Unified Realm and Hans von Ohain in Germany during the 1930s denoted the start of the stream age. Fly motors, powered by lamp oil based flight fuel, offered a few benefits over cylinder motors, including higher effectiveness, more noteworthy push, and the capacity to work at higher elevations.

The Second Great War assumed a significant part in propelling plane innovation, as both the Unified and Pivot powers created stream controlled airplane for military purposes. The Messerschmitt Me 262, a German fly contender, and the English Gloster Meteor were among the principal functional stream fueled airplane. These mechanical headways during the conflict established the groundwork for post-war business aeronautics and added to the quick extension of the avionics business.

The Post-War Aeronautics Blast and the Fly Age

The finish of The Second Great War got a flood of interest business flight and a developing longing for quicker, more effective air travel. The flying business saw a time of extraordinary development, with carriers, makers, and legislatures putting resources into the improvement of business fly carriers. Key improvements in this period incorporated the Boeing 707, the Douglas DC-8, and the de Havilland Comet.

These early stream carriers were controlled by turbofan motors, which further superior eco-friendliness and reach. Stream motors depended on flight lamp oil, normally alluded to as fly fuel, which offered the important energy thickness and ignition attributes for rapid air travel. The accessibility and moderateness of fly fuel made significant distance air go more open to a more extensive scope of travelers.

The Boeing 707, presented in the last part of the 1950s, denoted a critical achievement in business flying. It was the main effective stream aircraft intended for long stretch flights and had the reach to associate significant urban communities all over the planet. The Douglas DC-8 and the de Havilland Comet additionally added to the extension of worldwide air travel, interfacing landmasses and cultivating globalization.

The progress to the fly age reformed the aircraft business. With fly carriers fit for flying at higher velocities and heights, long stretch courses turned out to be financially suitable. Carriers could now offer cross-country and intercontinental flights, extending their organizations and taking special care of a developing interest for worldwide travel.

The effect of oil on the avionics business was twofold. To start with, the accessibility of stream fuel empowered the advancement of additional strong and effective motors, supporting the development of the avionics business. Second, the accommodation and productivity of stream controlled air head out prompted an expansion sought after for aeronautics fuel, further driving the extension of the oil business.

Air Travel as an Impetus for Monetary Development

The development of the flying business had sweeping monetary results. The development of business flying added to monetary turn of events, work creation, and expanded exchange and the travel industry. Here are a portion of the critical manners by which flying energized financial development:

Work Creation: The flying business made an extensive variety of business valuable open doors, from pilots and airline stewards to mechanics, designers, and ground staff. Furthermore, air terminals and flight related organizations, like upkeep and catering administrations, produced positions and financial movement.

The travel industry and Friendliness: The development of air travel opened up new open doors for the travel industry and cordiality enterprises. Vacationer locations all over the planet saw a convergence of guests, and inns, eateries, and vacation spots flourished. The capacity to travel effectively and rapidly via air helped support global the travel industry.

Exchange and Business: The avionics business worked with the fast development of products, associating makers and buyers in various areas of the world. Air freight administrations assumed a basic part in the globalization of exchange, empowering the transportation of high-esteem, time-touchy products.

Framework Advancement: The extension of the flying business prodded interests in air terminal foundation, including the development of new terminals, runways, and control towers. These ventures emphatically affected nearby economies and made positions in development and related enterprises.

Mechanical Progressions: The flight business' interest for state of the art innovation prompted headways in aeronautics, materials science, and security frameworks. These developments frequently tracked down applications in different enterprises, cultivating mechanical advancement and financial development.

The Job of Oil Stores and International Elements

The flying business' dependence on oil reached out to its requirement for a steady and secure inventory of stream fuel. As the business developed, so did its reliance on oil-creating locales all over the planet. The solidness of these locales and the international elements of oil-creating countries was the fate of central significance to the flight business.

Oil-delivering countries in the Center East, like Saudi Arabia, Iran, and Iraq, held huge stores of raw petroleum. The capacity to tie down admittance to these stores and keep up with stable stockpile chains was a basic consider guaranteeing the continuous development of the flight business. The advancement of long stretch flights and worldwide network made these locales decisively significant for flying and energy security.

International pressures and clashes in oil-creating areas could upset the flying business. The 1970s saw the effect of such struggles on oil costs and supply, with occasions like the Middle Easterner oil ban and the Iranian Unrest prompting huge changes in stream fuel costs. The business had to adjust to these provokes and investigate techniques to upgrade energy security.

Endeavors were made to enhance wellsprings of oil and foster vital stores to moderate the effect of supply interruptions. The Global Energy Organization (IEA) was laid out because of the oil emergencies of the 1970s, with an essential spotlight on planning crisis reaction gauges and advancing energy security.

The monetary effect of oil cost instability on the aeronautics business was mind boggling. While rising fuel expenses could strain carriers' productivity, they could likewise boost the advancement of additional eco-friendly airplane and functional practices. These difficulties incited innovative work endeavors pointed toward further developing eco-friendliness and diminishing the natural impression of avionics.

The Natural Test

While the flight business received critical financial rewards from its association with the oil business, it likewise confronted developing worries about its natural effect. The ignition of stream fuel in airplane motors brings about the arrival of ozone depleting substances and different contaminations that add to environmental change and air quality issues.

The natural effect of flying has turned into a subject of worldwide concern. The flying business, controllers, and legislatures have done whatever it may take to address these difficulties and lessen the area's carbon impression. Drives, for example, the reception of more eco-friendly airplane, the advancement of maintainable avionics powers (SAFs), and enhancements in air traffic the board have been sought after to moderate the natural effect of air travel.

The mission for more maintainable aeronautics has additionally prompted headways in airplane plan and impetus advances. Advancements like calmer, more eco-friendly motors, further developed streamlined features, and the investigation of electric and cross breed drive can possibly change the aeronautics business in the next few decades.

The Job of Oil in Military Flight

Military flight plays had a huge impact in molding the innovation and capacities of the flying business. Oil has been an essential asset in this unique situation, driving military airplane and supporting the vital and strategic goals of military all over the planet.

The Second Great War denoted the presentation of military avionics, with biplanes and triplanes utilized for observation, elevated battle, and besieging missions. These early airplanes were controlled by cylinder motors and normally utilized flight gas or other specific energizes.

The Second Great War achieved critical headways in military aeronautics innovation. The contention saw the far and wide utilization of military aircraft, planes, and transport airplane fueled by gas powered motors. These motors depended on aviation.

3.3 Oil's impact on the global shipping and transportation networks.

The worldwide delivery and transportation networks act as the conduits of the cutting edge world, working with the development of merchandise and individuals on an extraordinary scale. Key to the activity of these organizations is the job of oil, a flexible and energy-thick fuel that powers different methods of transportation, from freight boats to trucks and trains. This paper investigates the significant effect of oil on worldwide delivery and transportation organizations, diving into how the accessibility and dependence on oil have molded exchange, economies, and the manner in which we associate and trade merchandise across the world.

Sea Delivery: An Impetus for Worldwide Exchange

The transportation business is the foundation of global exchange, empowering the development of products across seas and between mainlands. The reception of oil as an essential fuel source assumed a critical part in changing oceanic transportation and growing worldwide exchange.

The progress from coal to oil in the sea business changed the proficiency of boats. Oil offered a few benefits over coal, including higher energy thickness, decreased discharges, and improved on refueling processes. Steam motors, energized by coal, were supplanted by more productive diesel motors that pre-owned oil as their essential fuel source.

The improvement of oil-fueled ships took into account longer journeys without the requirement for continuous refueling. This expanded effectiveness and reach empowered the worldwide development of products on a remarkable scale. Huge oil big haulers, intended to ship immense amounts of unrefined petroleum and oil based commodities, further worked with worldwide exchange by productively conveying oil from delivering districts to far off business sectors.

The accessibility of oil as a fuel source changed how products are moved. Freight ships, compartment vessels, and mass transporters fueled by oil motors made it financially savvy to get enormous amounts of merchandise across seas. Ports and harbors created broad framework to oblige these vessels, making them key hubs in the worldwide transportation organization.

The Panama Trench, a basic oceanic passage, went through huge updates and developments in the mid twentieth hundred years to oblige bigger vessels, including oil big haulers. The channel's significance in working with worldwide exchange was highlighted by its job in associating the Atlantic and Pacific Seas, diminishing travel times and transportation costs. The trench's modernization further sped up the globalization of the oceanic business and the development of merchandise.

The worldwide delivery industry's dependence on oil reaches out past impetus. Ships are controlled by oil-based fills, and many convey cargoes of raw petroleum and oil based commodities. The connection between oil creation and oceanic delivery is a commonly useful one, as the interest for oil drives the development of the transportation business, and transportation, thus, upholds the worldwide development of oil.

Difficulties and Open doors in Oceanic Transportation

While the sea delivering industry has encountered amazing development and mechanical progressions, it has additionally confronted a few difficulties. Natural worries, wellbeing guidelines, and international variables have formed the business' development and the job of oil inside it.

Ecological Worries: The utilization of oil-based energizes in transportation has raised worries about air contamination and ozone depleting substance emanations. Subsequently, there has been a push for cleaner and more productive impetus frameworks. Delivering organizations are investigating elective energizes, like melted gaseous petrol (LNG) and marine diesel oil, to diminish discharges and meet ecological guidelines.

Security Guidelines: The oceanic business is dependent upon severe wellbeing guidelines, including measures to forestall oil slicks and safeguard marine environments. Guidelines like the Global Show for the Anticipation of Contamination from Boats (MARPOL) have been executed to alleviate the natural effect of oil transport. These guidelines impact the plan and activity of oil big haulers and freight ships.

International Factors: The sea transporting industry is delicate to international turns of events, as delivery courses and admittance to key ports are dependent upon the political elements of beach front countries. International pressures and clashes can upset transportation paths and effect the development of oil and different products. Subsequently, industry partners should consider international dangers and attempt to guarantee the security and strength of worldwide shipping lanes.

Mechanical Progressions: The oceanic business is embracing innovative advancements to further develop proficiency and manageability. Shipowners are putting resources into more eco-friendly vessel plans, high level drive frameworks, and computerized innovations to advance courses and freight the executives. The reception of cutting edge route frameworks and robotization is improving wellbeing and functional productivity.

The development of worldwide exchange, the rising interest for shopper merchandise, and the development of online business have additionally raised the significance of oceanic transportation. The business' capacity to adjust to changing buyer inclinations and convey merchandise in a convenient and financially savvy way is crucial for supporting monetary development and global exchange.

Land Transportation: The Foundation of Territorial and Neighborhood Exchange

Land transportation organizations, including street, rail, and pipelines, are basic to territorial and neighborhood exchange, guaranteeing the productive development of merchandise inside and between nations. Oil plays had a principal impact in fueling land transportation, changing the coordinated operations and store network enterprises.

Street Transportation: The far reaching reception of the vehicle in the mid twentieth century changed individual portability and street transportation. Individual vehicles and business trucks depend on gas and diesel fuel, both got from oil, to drive gas powered motors. The comfort and adaptability of street transportation made it an essential mode for the development of products and individuals.

Shipping organizations, empowered by the accessibility and reasonableness of diesel fuel, changed the planned operations industry. Trucks turned into the essential mode for shipping products over short and significant distances, associating makers, wholesalers, and retailers. The shipping business' extension plays had a significant impact in store network the executives, giving effective and dependable transportation administrations.

The development of street transportation has prodded interests in street foundation, like thruways, scaffolds, and passages. States and confidential area substances have attempted to upgrade the quality and limit of street organizations, guaranteeing the proficient development of merchandise and supporting monetary turn of events.

Rail Transportation: Rail transportation has been a foundation of land-based exchange, giving proficient and practical choices for moving mass products over significant distances. Diesel trains, which supplanted steam motors, offered superior effectiveness, decreased working expenses, and expanded unwavering quality. This progress permitted railways to give practical transportation benefits and encouraged monetary advancement by working with the development of products over significant distances.

Oil likewise assumes a fundamental part in rail transportation, fueling the two trains and the creation of the steel and different materials utilized in rail foundation. Cargo trains, which transport merchandise going from natural substances to completed items, are a basic piece of the inventory network and the development of assets and wares.

Pipelines: Pipelines have arisen as a critical part of the transportation organization, explicitly for the development of oil and petroleum gas. Pipelines offer a protected and proficient method for shipping energy assets over significant distances, frequently from creation regions to treatment facilities and dissemination focuses.

The accessibility of pipelines has been instrumental in supporting the oil business' development and the proficient development of oil assets. Raw petroleum pipelines, for instance, transport oil from boring locales to processing plants, guaranteeing a consistent stockpile of oil based goods for transportation and different enterprises.

Chapter 4

Beyond Fuel: Petrochemicals and Plastics

The petrochemical business is a fundamental piece of our advanced world, and its impact arrives at a long ways past the gas we put in our vehicles. While the essential relationship with petrol is many times fuel, the creation and utilization of petrochemicals and plastics significantly affect our lives, economies, and the climate. In this talk, we will dig into the complicated trap of petrochemicals and plastics, investigating their importance, creation processes, ecological ramifications, and expected choices in a world that is progressively centered around supportability.

Petrochemicals, as the name recommends, are substance compounds got from petrol, a normally happening non-renewable energy source. These mixtures are the structure blocks for a great many items, from plastics and manufactured strands to drugs and manures. The petrochemical business' impact stretches out to different areas, including horticulture, medical services, car, development, and hardware. As the interest for these items keeps on filling in a globalized and industrialized world, the petrochemical business has become one of the most essential parts of the worldwide economy.

One of the central members in the petrochemical business is ethylene, a hydrocarbon gas created principally from flammable gas and oil. Ethylene fills in as the establishment for the development of different polymers, including polyethylene, which is the most broadly involved plastic on the planet.

The petrochemical business' ability to deliver ethylene is enormous, and it frequently outperforms the limit of even the oil refining industry. This embodies the complex connection between the oil and petrochemical areas, which frequently work pair.

Polyethylene, with its assorted scope of uses, has altered various ventures. From plastic sacks to clinical gadgets, its versatility has made it a foundation of current life. In any case, this pervasiveness has included some significant downfalls. The creation and removal of polyethylene and different plastics present critical natural difficulties.

The natural effect of plastics is profoundly disturbing. Plastics can require many years to break down in the regular habitat, prompting unavoidable contamination and natural surroundings annihilation. Marine life, specifically, experiences the plastic plague, as the world's seas are loaded up with plastic waste that compromises marine environments and even enters the human pecking order. The greatness of the issue is faltering, with a large number of lots of plastic waste produced every year.

While the ecological issues encompassing plastics are broadly perceived, the petrochemical business' job in worsening these worries is frequently underrated. The development of plastics is energy-concentrated and produces huge ozone depleting substance emanations, adding to environmental change. Moreover, the extraction and refinement of the petroleum derivatives fundamental for petrochemical creation have their own environmental effects, including living space annihilation, water contamination, and air contamination.

As we go up against the difficulties of environmental change and natural debasement, it is basic to analyze the petrochemical business' future and investigate choices. Progressing away from a weighty dependence on petrochemicals is a mind boggling try, as these mixtures are profoundly imbued in our day to day routines. Regardless, a few methodologies can assist with relieving the natural effect of petrochemicals and plastics.

One road for change is the advancement of bioplastics, which are gotten from sustainable assets like plants, green growth, and microscopic organisms. Bioplastics can be intended to be biodegradable, decreasing their natural impression. While bioplastics stand out and venture, they actually face difficulties concerning versatility, cost-viability, and execution contrasted with customary plastics.

Another procedure is the round economy, which means to limit squander and amplify asset proficiency. This approach includes reusing, reusing, and lessening the utilization of plastics. It requires changes in item plan, assortment frameworks, and shopper conduct. Accomplishing a round economy for plastics is a complicated errand that includes coordinated effort between legislatures, enterprises, and purchasers.

Substance reusing is an arising innovation that can possibly change the petrochemical business. Not at all like customary mechanical reusing, which includes dissolving and going back over plastics, synthetic reusing separates plastics into their constituent monomers or different synthetic substances. These can then be utilized to deliver new plastics or other significant materials. While substance reusing holds guarantee, it faces specialized, financial, and administrative provokes that should be defeated for broad reception.

Quite possibly of the main change in the petrochemical business is the shift towards sustainable and elective feedstocks. Customarily, petrochemicals are gotten from petroleum products, yet there is developing interest in utilizing biomass, carbon dioxide, and other feasible sources as unrefined components. This change is in accordance with the more extensive move towards a low-carbon, round economy.

The job of government arrangements and guidelines can't be put into words in that frame of mind in the petrochemical business. Legislatures can boost the reception of supportable practices, force expenses or requires on non-recyclable or non-biodegradable plastics, and set focuses for diminishing ozone harming substance outflows from petrochemical creation.

In the confidential area, organizations are perceiving the need to address the natural effect of their items and cycles. Many are putting resources into innovative work to track down feasible options in contrast to customary petrochemicals. Purchaser interest for eco-accommodating items is likewise pushing organizations to make changes in their stockpile chains.

The worldwide petrochemical industry is going through a change, driven by a mix of market elements, natural worries, and innovative progressions. This change, be that as it may, isn't without its difficulties and intricacies. The progress to more economical petrochemical rehearses requires a comprehensive methodology that envelops mechanical development, strategy changes, and changes in customer conduct.

One of the key difficulties confronting the petrochemical business is the requirement for more effective and manageable creation techniques. Customary petrochemical processes are energy-escalated and produce huge ozone harming substance emanations. To address this, innovative work endeavors are centered around growing all the more harmless to the ecosystem creation techniques, including the utilization of environmentally friendly power sources, carbon catch and usage, and further developed process productivity.

Mechanical headways are additionally driving development in the petrochemical business. The improvement of new impetuses, materials, and cycle advances is empowering the development of petrochemicals with less natural effects. For instance, the utilization of cutting edge impetuses can upgrade the selectivity and productivity of synthetic responses, decreasing waste and energy utilization.

The rising interest in zap and environmentally friendly power sources can possibly reshape the petrochemical business. As sustainable power turns out to be more open and reasonable, it could give a cleaner wellspring of energy for petrochemical processes, lessening their natural impression.

As well as tending to creation strategies, the petrochemical business is investigating elective feedstocks. Biomass, carbon dioxide, and modern waste streams are being considered as wellsprings of carbon for petrochemical creation. These elective feedstocks enjoy the benefit of being inexhaustible and possibly diminishing the business' dependence on non-renewable energy sources.

One of the eminent patterns in the petrochemical business is the ascent of bioplastics. These are plastics produced using sustainable sources, like cornstarch, sugarcane, or plant oils. Bioplastics stand out enough to be noticed for their capability to lessen the natural effect of plastics. They can be biodegradable or compostable, tending to a portion of the issues related with customary plastics.

Bioplastics have tracked down applications in different businesses, from bundling and farming to auto and customer merchandise. Notwithstanding, challenges remain, including the adaptability of creation, cost-viability, and the requirement for clear naming to recognize bioplastics from ordinary plastics.

The roundabout economy idea is one more road of progress in the petrochemical business. It advances decreasing waste and expanding asset effectiveness by reusing, reusing, and lessening the utilization of materials.

4.1 The role of oil in the development of petrochemicals.

The petrochemical business, a foundation of present day assembling and day to day existence, depends vigorously on oil as an essential feedstock. The expression "petrochemical" itself passes the business' personal association on to oil, and its development and advancement over the course of the years have been complicatedly connected with the accessibility and usage of oil. In this talk, we will investigate the essential job of oil in the improvement of petrochemicals, following the set of experiences, effect, and difficulties that have molded this crucial area.

The petrochemical business' underlying foundations can be followed back to the late nineteenth century when the oil business was in its early stages. Around then, raw petroleum was essentially esteemed for its utilization as a wellspring of enlightenment through lamp oil, which supplanted not so much productive but rather more hazardous lighting sources like whale oil and coal gas. The disclosure of new oilfields in districts like Pennsylvania, Texas, and California denoted the start of a worldwide shift towards petrol based energy sources.

As the oil business extended, so did the assortment of results separated from raw petroleum. Early treatment facilities essentially centered around refining unrefined petroleum to deliver lamp fuel, however they before long understood the capability of different hydrocarbons got from this plentiful asset. A side-effect of lamp fuel creation, gas, acquired noticeable quality with the ascent of the car business in the mid twentieth hundred years, exhibiting how the flexibility of oil was immediately perceived and bridled for different purposes.

The primary significant jump in the advancement of petrochemicals can be credited to the rise of the engineered elastic industry during The Second Great War. With regular elastic supplies from Southeast Asia disturbed by the contention, the interest for choices prompted the improvement of manufactured elastic utilizing oil-inferred feedstocks. This development denoted a huge achievement in the petrochemical business, growing its applications past fuel and illuminants.

The Second Great War additionally sped up the development of petrochemicals as the interest for engineered materials, including plastics and manufactured elastic, arrived at new levels. These materials were basic for the conflict exertion, and their significance in day to day existence kept on developing post-war. Plastics, specifically, started to track down their direction into many items, from bundling and materials to hardware and development materials. The large scale manufacturing of plastics turned into a main trait of the post-war time.

Oil, with its rich hydrocarbon content, arose as the essential wellspring of unrefined components for the petrochemical business. The hydrocarbons in raw petroleum, when appropriately handled, can be changed over into an extensive variety of petrochemical compounds. Ethylene, propylene, and butadiene, among others, are a portion of the fundamental structure blocks got from oil that act as forerunners for a horde of items.

Ethylene, specifically, assumes a crucial part in the petrochemical business. It is perhaps of the main feedstock and is acquired principally from petroleum gas and oil. Ethylene fills in as the establishment for the creation of different polymers, including polyethylene, which is the most broadly involved plastic on the planet. The cozy connection between the oil and petrochemical areas is clear in the way that ethylene creation limit frequently outperforms the limit of oil refining. This co-dependence between the two enterprises features how oil has powered the extension and expansion of the petrochemical area.

The broad utilization of oil in the petrochemical business has added to the development of economies and the change of social orders around the world. The accessibility of reasonable and flexible petrochemical items has worked with the improvement of different areas, including agribusiness, medical services, auto, development, and gadgets. Petrochemicals have empowered developments in various fields, prompting the formation of additional proficient and practical items.

Plastics, specifically, significantly affect different parts of day to day existence. From expendable bundling to clinical gadgets, plastic items have upset enterprises and worked on the personal satisfaction for billions of individuals. The lightweight and strong nature of plastics, joined with their flexibility and minimal expense, make them key in numerous applications.

The petrochemical business' prosperity, powered by the accessibility of oil as a feedstock, has not been without difficulties and contentions. One of the most squeezing concerns is the natural effect of plastics. Plastics, especially single-use plastics, have turned into a significant wellspring of contamination, stopping up landfills and dirtying the world's seas. The sluggish decay of plastics in the climate, which can require many years, presents huge natural and biological dangers.

Marine life, specifically, is helpless against plastic contamination, as tremendous amounts of plastic waste enter the world's seas, hurting marine biological systems and in any event, entering the human pecking order. Microplastics, little plastic particles, have been tracked down in different marine species and are a developing worry because of their potential wellbeing influences on people and natural life.

The creation of plastics itself is an asset escalated and energy-consuming cycle. It creates significant ozone harming substance outflows, adding to environmental change. Furthermore, the extraction and refinement of the petroleum derivatives important for petrochemical creation accompany their own arrangement of environmental issues, including natural surroundings annihilation, water contamination, and air contamination.

While the natural issues related with plastics are broadly perceived, the petrochemical business' part in fueling these worries is frequently underrated. The business' extension is vigorously dependent on the accessibility of oil and flammable gas, which are limited and non-sustainable assets. This over-dependence on non-renewable energy sources adds to ecological debasement as well as makes the petrochemical area helpless against changes in oil costs and supply disturbances.

The ecological difficulties related with petrochemicals and plastics have prompted developing tension on states, enterprises, and purchasers to look for additional reasonable other options and practices. While the progress away from petrochemicals and plastics is mind boggling, a few methodologies can assist with relieving the ecological effect of these businesses.

One methodology is the advancement of bioplastics, which are gotten from inexhaustible assets like plants, green growth, and microscopic organisms. Bioplastics can be intended to be biodegradable, lessening their ecological impression. They offer the possibility to address a portion of the issues related with conventional plastics, like their diligence in the climate.

Bioplastics are acquiring consideration and speculation, with organizations and specialists attempting to work on their presentation, cost-viability, and versatility. Nonetheless, challenges remain, including the need to recognize bioplastics from customary plastics and address worries about contest for agrarian assets.

The roundabout economy idea is one more system for lessening the ecological effect of petrochemicals and plastics. It centers around limiting waste and boosting asset proficiency through practices like reusing, reusing, and lessening utilization. Accomplishing a roundabout economy for plastics requires changes in item plan, assortment frameworks, and customer conduct, as well as interest in reusing foundation.

Synthetic reusing, an arising innovation in the petrochemical business, holds guarantee for tending to a portion of the ecological difficulties. Dissimilar to customary mechanical reusing, which includes liquefying and going back over plastics, substance reusing separates plastics into their constituent monomers or different synthetic compounds. These can then be utilized to create new plastics or other important materials. While compound reusing faces specialized, monetary, and administrative difficulties, it can possibly change the business.

Perhaps of the main change in the petrochemical business is the shift towards sustainable and elective feedstocks. While customary petrochemicals are gotten from petroleum products, there is developing interest in utilizing biomass, carbon dioxide, and other reasonable sources as unrefined substances. This progress lines up with the more extensive move towards a low-carbon, round economy and offers the possibility to decrease the business' dependence on limited fossil assets.

Government strategies and guidelines assume a basic part in driving change in the petrochemical business. Legislatures can boost the reception of manageable practices, force charges or demands on non-recyclable or non-biodegradable plastics, and set focuses for decreasing ozone depleting substance outflows from petrochemical creation. Administrative structures can make a level battleground for reasonable other options and support all inclusive reception of eco-accommodating practices.

The confidential area is likewise effectively answering the natural difficulties presented by the petrochemical business. Organizations are perceiving the need to address the ecological effect of their items and cycles. Many are putting resources into innovative work to track down maintainable options in contrast to customary petrochemicals. Buyer interest for eco-accommodating items is additionally pushing organizations to make changes in their stockpile chains and item contributions.

The worldwide petrochemical industry is going through a change, driven by a mix of market elements, ecological worries, and innovative headways. This change is mind boggling and diverse, as it includes changes underway techniques, feedstocks, and buyer conduct.

Endeavors to make petrochemical creation all the more harmless to the ecosystem are continuous, with an emphasis on energy proficiency, decreased emanations, and manageable practices. Mechanical developments, like promotion.

4.2 The birth of the plastics industry.

The introduction of the plastics business is an account of development, need, and the steady quest for new materials that changed our reality. Plastics, as we probably are aware them today, have become universal in our regular routines, however their starting point can be followed back to the nineteenth century when researchers and designers started exploring different avenues regarding new manufactured materials. In this story, we will investigate the captivating excursion that prompted the production of plastics, the different kinds of plastics that arose, and the significant effect they have had on enterprises, economies, and social orders.

The narrative of plastics starts with the quest for materials that could imitate the properties of normal assets, like ivory, horn, and shellac. These materials were valued for their sturdiness and flexibility yet were restricted in supply and frequently expensive. The requirement for choices that could be made for a bigger scope and at a lower cost led to the improvement of early plastics.

One of the earliest manufactured polymers to acquire noticeable quality was celluloid. In 1862, British bloke Alexander Parkes presented Parkesine, which was gotten from cellulose, a characteristic polymer found in plant cell walls. Parkesine was an earth shattering development, as it very well may be shaped when warmed and was at first utilized for different applications, including buttons, brushes, and enhancing things.

Parkesine, in any case, was not without its limits. It was profoundly combustible and inclined to breaking. This prompted further trial and error and the possible advancement of celluloid by American creator John Wesley Hyatt during the 1870s. Celluloid, which likewise involved cellulose as its essential fixing, was not so much combustible but rather more strong than Parkesine. It tracked down a great many purposes, including as a substitute for ivory in billiard balls, as well as in photography and entertainment worlds for the creation of film reels.

While celluloid was a huge forward-moving step in the development of plastics, the genuine birth of the plastics business can be credited to Leo Baekeland's creation of Bakelite in 1907. Bakelite, frequently thought to be the main genuine plastic, was a phenol-formaldehyde tar. Not at all like prior plastics, Bakelite was completely engineered, and it showed wonderful properties, like high intensity opposition, electrical protecting capacities, and the capacity to be formed into complex shapes.

Bakelite's prosperity altered the plastics business, as it tended to various requirements in different applications. It was utilized widely in electrical and auto enterprises, where its electrical protecting properties and intensity obstruction were important. Bakelite additionally tracked down applications in buyer products, like radios, phones, and kitchenware.

Bakelite's adaptability propelled further advancement in the plastics business. The 1920s and 1930s saw the improvement of various manufactured plastics, each with its exceptional properties and applications. For instance, polystyrene was presented during the 1930s, and its straightforwardness, inflexibility, and simplicity of trim made it ideal for things like straightforward bundling, optical gadgets, and dispensable flatware.

The presentation of polyvinyl chloride (PVC) during the 1920s carried an adaptable plastic with different purposes. PVC is generally used today in the development business for pipes, electrical protection, and vinyl siding. During a similar period, polyethylene was found, offering a lightweight, adaptable plastic with phenomenal electrical protecting properties. It turned into the establishment for different plastic items, from plastic sacks to clinical gadgets.

Nylon, designed in the last part of the 1930s by Wallace Carothers at DuPont, was a significant jump forward. Nylon was the primary engineered fiber, and its solidarity and flexibility prompted its utilization in materials, ropes, and at last, the production of nylon stockings, which turned into a sensation. The Second Great War additionally highlighted the significance of plastics, as they assumed a crucial part in military applications, including airplane parts and defensive stuff.

The mid-twentieth century saw the advancement of various plastics that would proceed to change different ventures. High-thickness polyethylene (HDPE) was presented, bringing a plastic known for its solidarity, sturdiness, and synthetic opposition. It tracked down applications in bundling, development, and, surprisingly, as the material for the notable plastic milk container.

Polypropylene, presented during the 1950s, offered a flexible plastic with great synthetic opposition and strength. It has since turned into a staple in the bundling business and is utilized for items going from jugs to food compartments. Moreover, polycarbonate, known for its effect opposition and optical clearness, found applications in eyeglass focal points, security safeguards, and in the long run, minimized plates and DVDs.

The coming of the 21st century has seen a proceeded with flood in plastics development. Architects and researchers have created progressed designing plastics, for example, polyimides and fluid precious stone polymers, which offer high-temperature opposition and remarkable mechanical properties. These materials are utilized in aviation, car, and gadgets businesses, where outrageous execution necessities exist.

Another imperative improvement is the ascent of bioplastics. These are plastics gotten from inexhaustible assets, like cornstarch, sugarcane, or plant oils.

Bioplastics offer a more supportable option in contrast to customary plastics, and they are frequently biodegradable or compostable. This advancement is in accordance with the developing worldwide consciousness of natural issues and the need to decrease the ecological effect of plastics.

Notwithstanding these developments, added substance assembling or 3D printing has given another outskirts to plastics. This innovation takes into consideration the exact layer-by-layer development of mind boggling objects, utilizing different plastics and composite materials. It can possibly alter fabricating, taking into consideration modified items and more proficient creation processes.

The plastics business has progressed significantly since its beginning, with a rich history of creation, improvement, and change. Plastics have become basic in our day to day routines, and their adaptability has driven their utilization in essentially every industry, from medical care and bundling to aviation and hardware.

The progress of plastics can be credited to their exceptional properties, including their capacity to be formed into essentially any shape, their sturdiness, and their lightweight nature. These properties, joined with the capacity to deliver plastics for scale and moderately minimal price, have made them fundamental materials in the cutting edge world.

The plastics business' effect on economies and enterprises has been significant. It has empowered the large scale manufacturing of shopper merchandise, changed bundling and transportation, and worked with mechanical headways. Plastics play had a basic impact in the improvement of items like PCs, clinical gadgets, and broadcast communications gear.

In any case, the outcome of plastics has not come without difficulties and concerns. One of the most major problems is plastic contamination. Plastics, particularly single-use plastics, have turned into a significant wellspring of ecological contamination. Deficient removal and reusing rehearses have prompted plastic waste collecting in landfills, seas, and regular territories, actually hurting natural life and biological systems.

Marine life is especially defenseless against plastic contamination, as plastic flotsam and jetsam in the seas can ensnare and hurt creatures, or be ingested, prompting injury or passing. Microplastics, minuscule plastic particles, have been tracked down in different marine species and are a developing worry because of their potential wellbeing influences on people and untamed life.

The natural effect of plastics stretches out past contamination. The development of plastics is energy-serious and creates critical ozone harming substance emanations, adding to environmental change. Furthermore, the extraction and refinement of the petroleum derivatives fundamental for plastic creation have their own environmental effects, including natural surroundings obliteration, water contamination, and air contamination.

As the world stands up to the difficulties of environmental change and ecological debasement, the plastics business is under expanding examination, and endeavors are in progress to address its natural effect. The advancement of bioplastics, which are gotten from sustainable assets and are frequently biodegradable, is one road to diminish the ecological impression of plastics.

The roundabout economy idea advances reusing and asset productivity, meaning to limit squander and boost the utilization of materials. Accomplishing a roundabout economy for plastics includes changes in item plan, assortment frameworks, and purchaser conduct, as well as interest in reusing foundation.

Synthetic reusing, an arising innovation, holds the possibility to change the plastics business. Not at all like conventional mechanical reusing, which includes liquefying and going back over plastics, substance reusing separates plastics into their constituent monomers or different synthetic compounds. These can then be utilized to deliver new plastics or other important materials. While synthetic reusing faces specialized, financial, and administrative difficulties, it can possibly essentially decrease plastic waste and ecological effect.

Another remarkable improvement is the shift towards sustainable and elective feedstocks for plastics. Customarily, plastics are gotten from petroleum derivatives, yet there is developing interest in utilizing biomass, carbon dioxide, and other supportable sources as natural substances. This change lines up with the more extensive move towards a low-carbon, roundabout economy.

Government strategies and guidelines are instrumental in driving change in the plastics business. States can boost the reception of manageable practices, force duties or tolls on non-recyclable or non-biodegradable plastics, and set focuses for decreasing ozone harming substance discharges from plastic.

4.3 The proliferation of synthetic materials in everyday life.

The multiplication of manufactured materials in regular daily existence is a demonstration of human resourcefulness and our capacity to make creative answers for fulfill the needs of a quickly impacting world. Manufactured materials, going from plastics to engineered filaments, have changed various ventures and have become essential to current living. In this conversation, we will dig into the sweeping effect of engineered materials on day to day existence, investigating their different applications, benefits, and the difficulties they present regarding manageability and ecological effect.

Engineered materials, by definition, are substances that are falsely made by synthetically adjusting or handling unrefined components. While they envelop many materials, plastics are maybe the most notorious and broadly perceived classification. Plastics, got from petrochemicals, have become universal in our day to day routines, utilized for a broad cluster of utilizations, from bundling to clinical gadgets, car parts to buyer hardware.

One of the critical benefits of plastics is their adaptability. They can be shaped into practically any shape, which makes them reasonable for an immense scope of items, from the adaptable, lightweight parts of a cell phone to the strong, solid development materials utilized in the structure business. Plastics have changed ventures by offering savvy and strong arrangements that have further developed item execution and diminished creation costs.

Engineered filaments are one more class of manufactured materials that have tracked down a huge spot in our lives. The improvement of engineered strands started with the production of nylon in the last part of the 1930s, trailed by the presentation of polyester, acrylic, and other manufactured filaments. These materials offer particular benefits over regular strands, like cotton or fleece. They are much of the time more solid, have better protection from ecological circumstances, and can be designed for explicit properties, for example, dampness wicking, fire opposition, and UV insurance.

Engineered strands are indispensable to the material business, where they are utilized to make attire, footwear, and other material items. They have upset the design business by empowering the development of an extensive variety of reasonable, agreeable, and classy dress. Moreover, manufactured strands have tracked down applications in the development of open air gear, including waterproof coats, climbing boots, and elite execution athletic apparel.

Moreover, the improvement of manufactured materials plays had a critical impact in the transportation business. Lightweight and strong manufactured composites have supplanted conventional materials like wood, metal, and normal strands in the development of vehicles, airplane, and marine vessels. This shift has added to upgraded eco-friendliness, decreased outflows, and further developed wellbeing norms. Carbon fiber-supported composites, for example, are broadly utilized in the aviation and auto areas because of their wonderful solidarity to-weight proportion.

Engineered materials have likewise made critical advances in the medical care area. The development of clinical gadgets and hardware, like needles, catheters, and prosthetics, depends on the flexibility and bio-compatibility of manufactured materials. Furthermore, manufactured polymers assume a pivotal part in drug conveyance frameworks, considering the controlled arrival of prescriptions to treat different ailments.

The development of the hardware business could never have been conceivable without manufactured materials. Protecting materials, conductive polymers, and lightweight parts have made ready for the advancement of more modest, more proficient electronic gadgets. Engineered materials empower the scaling down of parts and the assembling of adaptable presentations and lightweight batteries.

The multiplication of manufactured materials in day to day existence has not been restricted to a couple of enterprises; it has affected virtually every feature of present day living.

From family things like cooking wares and furniture to sporting hardware like surfboards and tennis rackets, engineered materials have become fundamental to our ordinary encounters. They offer arrangements that are in many cases more reasonable, solid, and adaptable than their normal partners.

While the upsides of manufactured materials are clear, their broad use has raised worries about natural maintainability and the drawn out influence in the world. The development of engineered materials, particularly plastics, is energy-serious and creates significant ozone depleting substance discharges. Also, the extraction and refinement of the petroleum derivatives fundamental for engineered material creation have their own natural effects, including living space obliteration, water contamination, and air contamination.

Plastic contamination, specifically, has arisen as a worldwide natural emergency. Plastics are non-biodegradable and can require many years to break down in the common habitat. Subsequently, plastic waste aggregates in landfills, seas, and normal environments, hurting natural life and biological systems. Marine life is especially defenseless against plastic contamination, as plastic flotsam and jetsam in the seas can snare and damage creatures or be ingested, prompting injury or demise. Microplastics, little plastic particles, have been tracked down in different marine species and are a developing worry because of their potential wellbeing influences on people and natural life.

Endeavors are in progress to address the ecological effect of engineered materials and advance more supportable practices. One methodology is the improvement of bioplastics, which are gotten from inexhaustible assets like plants, green growth, and microorganisms. Bioplastics can be intended to be biodegradable or compostable, diminishing their natural impression. This advancement is lined up with the developing world-wide consciousness of ecological issues and the need to diminish the natural effect of plastics.

The round economy idea has built up forward momentum as a procedure to lessen squander and expand asset effectiveness. It centers around reusing, reusing, and lessening the utilization of materials. Accomplishing a roundabout economy for engineered materials includes changes in item plan, assortment frameworks, and shopper conduct, as well as interest in reusing foundation. The objective is to limit squander and expand the life expectancy of materials, decreasing the requirement for virgin assets.

Compound reusing, an arising innovation, holds guarantee for tending to a portion of the natural difficulties related with engineered materials. Not at all like customary mechanical reusing, which includes dissolving and going back over materials, compound reusing separates materials into their constituent monomers or different synthetic substances. These can then be utilized to deliver new materials or other significant items. While compound reusing faces specialized, monetary, and administrative difficulties, it can possibly fundamentally lessen squander and ecological effect.

The shift towards sustainable and elective feedstocks is one more critical improvement in the journey for additional supportable engineered materials. Customarily, engineered materials are gotten from petroleum products, yet there is developing interest in utilizing biomass, carbon dioxide, and other supportable sources as unrefined components. This change lines up with the more extensive move towards a low-carbon, roundabout economy and offers the possibility to diminish the business' dependence on limited fossil assets.

Government arrangements and guidelines are basic drivers of progress in the manufactured materials industry. Legislatures can boost the reception of supportable practices, force charges or imposes on non-recyclable or non-biodegradable materials, and set focuses for lessening ozone depleting substance outflows from material creation. Administrative systems can make a level battleground for manageable other options and energize expansive reception of eco-accommodating practices.

The confidential area is additionally effectively answering the natural difficulties presented by manufactured materials. Many organizations are putting resources into innovative work to track down supportable options in contrast to conventional materials. Purchaser interest for eco-accommodating items is pushing organizations to make changes in their stockpile chains and item contributions. Supportability has turned into a critical driver for item development and promoting methodologies.

All in all, the multiplication of engineered materials in daily existence has changed ventures, economies, and social orders, giving flexible, strong, and practical answers for a large number of uses. Plastics and engineered strands have upset different areas, from bundling and medical services to auto and hardware, offering additional opportunities and driving innovative progressions.

Notwithstanding, the inescapable utilization of manufactured materials has raised worries about their natural effect and manageability. Plastic contamination, energy-escalated creation processes, and the dependence on petroleum products have featured the requirement for additional mindful practices. Endeavors to address these difficulties are in progress, with an emphasis on the improvement of bioplastics, the advancement of the roundabout economy, substance reusing, and the utilization of sustainable and elective feedstocks.

Government approaches and guidelines, as well as buyer interest for eco-accommodating items, are driving changes in the engineered materials industry. The confidential area is effectively putting resources into economical other options and more dependable inventory chains. The shift towards a more practical and naturally mindful future for manufactured materials is an intricate and continuous undertaking, mirroring our developing comprehension of the effect of these materials in the world.

5 |

Chapter 5

Geopolitics of Oil

Oil, frequently alluded to as the "soul of the worldwide economy," assumes a focal part in molding the international relations of the world. The extraction, creation, circulation, and utilization of oil have been critical elements affecting worldwide relations, security elements, and the monetary fortunes of countries. This conversation will investigate the multifaceted and frequently combative connection among oil and international affairs, looking at how the control and stream of oil have generally been a wellspring of force and struggle, the job of significant oil-delivering countries, and the developing difficulties and open doors in a world progressively centered around practical energy choices.

Verifiable Point of view: Oil as a Wellspring of Force and Struggle

The international significance of oil became apparent during the twentieth hundred years, as the interest for this imperative energy source flooded, especially with the ascent of the auto and the spread of industrialization. This expanded interest prompted critical disclosures of oil holds in different areas of the planet, making way for international moving and contest. The control of oil holds and admittance to world-wide oil markets became focal key contemplations for some countries.

The Center East, specifically, arose as a point of convergence of worldwide oil international relations. The locale contains a portion of the world's biggest demonstrated oil saves, and its political dependability has been fundamental to guaranteeing a predictable inventory of oil to worldwide business sectors. The foundation of the Association of the Petrol Trading Nations (OPEC) in 1960 by significant oil-delivering countries in the Center East set the district's impact over worldwide oil costs and supplies.

All through the twentieth 100 years, admittance to oil assets and the longing to control or get oil supplies prompted different global struggles and unions. The two Universal Conflicts were set apart by the essential significance of oil, with both the Partners and the Pivot powers perceiving the basic job of oil in fighting and monetary strength. The control of oil fields in districts like the Center East, Africa, and Southeast Asia turned into an essential goal in clashes, for example, the Bay Conflict, the Iran-Iraq War, and the attack of Kuwait by Iraq.

In the last 50% of the twentieth hundred years, the US and the Soviet Association took part in a worldwide battle for impact, frequently described as the Virus War. Oil-rich districts and countries were successive landmarks in this international battle, with the two superpowers trying to lay out unions and secure admittance to fundamental oil assets.

The International Job of Significant Oil-Creating Countries

A few significant oil-delivering countries play played key parts in molding the international scene of the oil business and worldwide energy elements. The accompanying countries have employed critical impact in such manner:

Saudi Arabia: As one of the world's biggest oil makers, Saudi Arabia has been a key part in worldwide energy legislative issues. The realm is an establishing individual from OPEC and has customarily assumed a focal part in setting worldwide oil costs by changing its creation levels. Saudi Arabia's immense oil holds and its situation as the world's top oil exporter have made it a central part in the international relations of oil.

Russia: Russia is one of the world's top oil makers and a critical player in worldwide energy markets. The country's political impact in the oil and gas area is enhanced by its command over key travel courses and framework, like pipelines. Russia's state-controlled energy goliath, Gazprom, and its oil organizations, Rosneft and Lukoil, are vital participants in the worldwide energy industry.

The US: The US is both a significant oil maker and shopper, and its approaches, procedures, and partnerships fundamentally influence worldwide energy elements. The U.S. has generally assumed a persuasive part in oil international relations through its associations with oil-rich countries, its job as a tactical underwriter of safety in key districts, and its conciliatory endeavors to settle oil markets.

Iran: Iran has significant oil holds and has generally assumed a vital part in the Center East's oil international affairs. Sanctions, local contentions, and political variables have added to Iran's intricate relationship with the worldwide energy industry. Iran has on occasion involved its oil as a political device, and its relations with Western countries have been impacted by these energy contemplations.

Venezuela: Venezuela has a portion of the world's biggest demonstrated oil saves, however its oil industry has confronted various difficulties, including monetary flimsiness, political unrest, and assents. The nation's oil international affairs have been set apart by vacillations underway, estimating questions, and territorial strains.

The International Contemplations of Oil Supply and Security

Admittance to dependable and secure supplies of oil is of vital significance for nations trying to keep up with energy security. International contemplations frequently rotate around guaranteeing a steady stockpile of oil, both locally and from unfamiliar sources. The accompanying key international elements impact oil supply and security:

Oil Travel Courses: The transportation of oil from maker countries to shopper markets depends on a perplexing organization of pipelines, big hauler courses, and oceanic chokepoints. The control and security of these travel courses have significant international ramifications. The

Waterway of Hormuz, for instance, is an indispensable oceanic entry through which a critical part of worldwide oil shipments passes. Disturbances or clashes here can prompt huge stock interruptions.

Political Security: The political soundness of oil-creating countries is a basic figure guaranteeing a steady oil supply. Political distress, upsets, nationwide conflicts, and endorses can all disturb oil creation and commodities, influencing worldwide oil markets.

Energy Unions and Security Arrangements: Numerous nations have gone into energy collusions and security arrangements to get their oil supplies. These arrangements frequently include safeguard settlements and shared help to safeguard energy framework and supply lines.

Value Unpredictability: The change of oil costs can have huge international repercussions. Lower oil costs can strain the economies of significant oil-creating countries, prompting political and social turmoil. On the other hand, higher oil costs can influence shopper economies and influence energy strategy choices.

Natural and Environment Concerns: Ecological and environment contemplations are progressively forming the international relations of oil. The worldwide change towards cleaner and more practical energy sources has prompted endeavors to lessen oil reliance and enhance energy portfolios. Countries are progressively forced to lessen their fossil fuel byproducts and relieve the natural effect of their energy approaches.

Difficulties and Potential open doors in the Period of Supportable Energy

The international affairs of oil are developing as the world appearances squeezing difficulties connected with environmental change and ecological manageability. The need to lessen ozone harming substance outflows and change to reasonable energy sources is reshaping the energy scene and the connections between oil-delivering and consuming countries.

Challenges:

Environment Concerns: The basic to battle environmental change and lessen fossil fuel byproducts is moving the energy scene. Countries are feeling the squeeze to embrace cleaner energy sources and decline their dependence on petroleum products, including oil.

Energy Progress: The change to sustainable power sources, for example, sun oriented, wind, and electric vehicles, is disturbing conventional energy markets. Oil-creating countries are confronted with the test of adjusting to changing interest and worldwide energy inclinations.

Monetary Effect: The decrease in oil request can have critical financial ramifications for oil-subordinate countries. The soundness of their economies might be compromised, prompting political and social difficulties.

International Realignment: The shift toward manageable energy sources is adjusting international connections. Countries with plentiful sustainable assets, similar to daylight and wind, are acquiring vital significance, while oil-creating countries might see a decrease in their impact.

5.1 The struggle for control of oil resources.

The worldwide journey for control of oil assets has been a characterizing component of the twentieth and 21st hundreds of years. Oil, frequently alluded to as "dark gold," is a crucial and limited asset that has energized economies, enterprises, and whole countries. The essential significance of oil has prompted international moving, clashes, collusions, and fights for control that keep on molding the world we live in today. This story will dig into the multi-layered battle for control of oil assets, looking at the authentic, monetary, and political elements of this continuous test.

Verifiable Viewpoint: Oil as an Impetus for Power

The cutting edge time of oil double-dealing started during the nineteenth century when Edwin Drake bored the principal business oil well in Titusville, Pennsylvania, in 1859. This noticeable the introduction of the petrol business and the start of an extraordinary power in international affairs.

The US was the origination of the worldwide oil industry, and its fast development as an oil maker in the late nineteenth and mid twentieth hundreds of years had significant ramifications for global illicit relationships. Oil-controlled enterprises and advancements like the gas powered motor and the car altered transportation and prompted expanded interest for oil.

In the mid twentieth 100 years, huge oil disclosures were made in different areas of the planet, quite in the Center East. These disclosures laid out the locale as a basic player in worldwide oil international relations. The control of oil holds and admittance to worldwide oil markets became focal vital contemplations for some countries.

The Center East's rise as a point of convergence of worldwide oil international relations had extensive ramifications. The district contains a portion of the world's biggest demonstrated oil holds, and its political solidness has been fundamental to guaranteeing a predictable stock of oil to worldwide business sectors.

The foundation of the Association of the Oil Trading Nations (OPEC) in 1960 by significant oil-delivering countries in the Center East hardened the area's impact over worldwide oil costs and supplies. OPEC turned into a strong entertainer in the worldwide energy industry, attempting to set oil creation shares and impact costs to help its part nations.

The battle for control of oil assets has generally elaborate significant abilities looking to tie down admittance to these fundamental stores. The two Universal Conflicts were set apart by the essential significance of oil, with both the Partners and the Pivot powers perceiving the basic job of oil in fighting and monetary strength. The control of oil fields in districts like the Center East, Africa, and Southeast Asia turned into a key goal in clashes, for example, the Bay Conflict, the Iran-Iraq War, and the intrusion of Kuwait by Iraq.

Oil was additionally vital to the opposition between the US and the Soviet Association during the Virus War. The two superpowers perceived the significance of tying down admittance to oil-rich districts, and their worldwide battles for impact frequently rotated around energy assets.

Key International Contemplations in the Battle for Oil

Oil Travel Courses: The transportation of oil from maker countries to buyer markets depends on a complicated organization of pipelines, big hauler courses, and oceanic chokepoints. The control and security of these travel courses have significant international ramifications. The Waterway of Hormuz, for instance, is an imperative oceanic entry through which a critical part of worldwide oil shipments passes. Disturbances or clashes here can prompt critical stockpile interruptions.

Political Soundness: The political solidness of oil-creating countries is a basic consider guaranteeing a predictable oil supply. Political turmoil, upsets, nationwide conflicts, and authorizes can all disturb oil creation and commodities, affecting worldwide oil markets.

Energy Partnerships and Security Arrangements: Numerous nations have gone into energy unions and security arrangements to get their oil supplies. These arrangements frequently include safeguard settlements and shared help to safeguard energy framework and supply lines.

Value Unpredictability: The vacillation of oil costs can have critical international repercussions. Lower oil costs can strain the economies of significant oil-creating countries, prompting political and social turmoil. Alternately, higher oil costs can influence purchaser economies and influence energy strategy choices.

Natural and Environment Concerns: Ecological and environment contemplations are progressively forming the international relations of oil. The worldwide change towards cleaner and more reasonable energy sources has prompted endeavors to lessen oil reliance and enhance energy portfolios. Countries are progressively constrained to diminish their fossil fuel byproducts and alleviate the natural effect of their energy arrangements.

The International Job of Significant Oil-Delivering Countries

A few significant oil-delivering countries play played key parts in molding the international scene of the oil business and worldwide energy elements. The accompanying countries have employed huge impact in such manner:

Saudi Arabia: As one of the world's biggest oil makers, Saudi Arabia has been a key part in worldwide energy legislative issues. The realm is an establishing individual from OPEC and has customarily assumed a focal part in setting worldwide oil costs by changing its creation levels. Saudi Arabia's huge oil holds and its situation as the world's top oil exporter have made it a key part in the international affairs of oil.

Russia: Russia is one of the world's top oil makers and a huge player in worldwide energy markets. The country's political impact in the oil and gas area is enhanced by its command over key travel courses and framework, like pipelines. Russia's state-controlled energy goliath, Gazprom, and its oil organizations, Rosneft and Lukoil, are key participants in the worldwide energy industry.

The US: The US is both a significant oil maker and customer, and its approaches, procedures, and unions essentially influence worldwide energy elements. The U.S. has generally assumed a persuasive part in oil international relations through its associations with oil-rich countries, its job as a tactical underwriter of safety in key districts, and its conciliatory endeavors to settle oil markets.

Iran: Iran has significant oil saves and has generally assumed a vital part in the Center East's oil international relations. Sanctions, territorial struggles, and political elements have added to Iran's mind boggling relationship with the worldwide energy industry. Iran has on occasion involved its oil as a political instrument, and its relations with Western countries have been impacted by these energy contemplations.

Venezuela: Venezuela has a portion of the world's biggest demonstrated oil saves, however its oil industry has confronted various difficulties, including monetary unsteadiness, political strife, and authorizations. The nation's oil international affairs have been set apart by vacillations underway, valuing debates, and provincial strains.

Difficulties and Amazing open doors in the Time of Practical Energy

The battle for control of oil assets is developing as the world countenances squeezing difficulties connected with environmental change and natural supportability. The need to lessen ozone depleting substance emanations and change to practical energy sources is reshaping the energy scene and the connections between oil-creating and consuming countries.

Challenges:

Environment Concerns: The basic to battle environmental change and diminish fossil fuel byproducts is moving the energy scene. Countries are feeling the squeeze to embrace cleaner energy sources and reduction their dependence on petroleum derivatives, including oil.

Energy Change: The progress to environmentally friendly power sources, for example, sunlight based, wind, and electric vehicles, is disturbing customary energy markets. Oil-delivering countries are confronted with the test of adjusting to changing interest and worldwide energy inclinations.

Financial Effect: The decrease in oil request can have huge monetary ramifications for oil-subordinate countries. The soundness of their economies might be undermined, prompting political and social difficulties.

International Realignment: The shift toward economical energy sources is changing international connections. Countries with plentiful sustainable assets, similar to daylight and wind, are acquiring key significance, while oil-delivering countries might see a decrease in their impact.

5.2 OPEC and the oil crises of the 20th century.

The Association of the Oil Trading Nations (OPEC) has assumed a focal and persuasive part in molding the worldwide oil industry and the international affairs of energy in the twentieth 100 years. OPEC, a consortium of significant oil-delivering countries, was established in 1960 to state command over their oil assets, oversee creation levels, and impact oil costs. All through the twentieth hundred years, OPEC's activities and choices had extensive results, finishing in the oil emergencies of the 1970s that significantly affected the worldwide economy, energy approaches, and global relations. This account investigates the historical backdrop of OPEC, its development, and its crucial job in the oil emergencies that characterized the twentieth 100 years.

The Introduction of OPEC

Shaping a worldwide association of oil-delivering countries flourished during the twentieth hundred years as these nations looked to declare command over their assets and secure a decent amount of the abundance produced by the oil business. The development of OPEC in 1960 denoted a critical stage toward this path.

The establishing individuals from OPEC included Iran, Iraq, Kuwait, Saudi Arabia, and Venezuela. These nations considered OPEC to be a way to all in all address their complaints with significant Western oil organizations, which had ruled the oil business and controlled evaluating and creation levels for quite a long time.

OPEC's expressed targets incorporated the accompanying:

Practicing sway over their oil assets.

Controlling creation and estimating to guarantee steady and fair profits from their oil.

Advancing solidarity and participation among part nations.

In the early long periods of its presence, OPEC confronted a difficult and frequently wild excursion. The association needed to explore the intricacies of its part countries' inclinations, coordinate creation shares, and face the force of significant Western oil organizations.

The 1960s: OPEC's Early stages

In its initial years, OPEC attempted to champion itself on the worldwide stage. The consortium confronted inside divisions and outside challenges. Western oil organizations opposed OPEC's endeavors to deal with evaluating and creation, and OPEC individuals were in many cases in conflict over creation levels and amounts. The association's power and impact were restricted, and it confronted wariness and opposition from laid out oil-delivering and consuming countries.

Toward the finish of the 1960s, OPEC had gained some headway in accomplishing its objectives, yet its effect on the worldwide oil market stayed restricted. It was during the 1970s that OPEC would ascend to noticeable quality and significantly influence worldwide energy elements.

The 1970s: The Oil Emergencies

The 1970s denoted a defining moment throughout the entire existence of OPEC and the worldwide energy scene. A progression of occasions, frequently alluded to as the oil emergencies, shook the world and featured OPEC's freshly discovered power. These emergencies were established in political, monetary, and international factors that would rethink worldwide energy elements.

The Main Oil Emergency (1973):

The main oil emergency started with the Yom Kippur Battle in October 1973, when an alliance of Bedouin states, including Saudi Arabia and other OPEC individuals, forced an oil ban on the US and other Western countries in light of their help for Israel during the contention. The ban prompted a huge decrease in oil supplies and set off a sharp expansion in oil costs.

The results of the ban were serious, as oil costs quadrupled in only a couple of months. This unexpected and emotional expansion in oil costs, combined with supply deficiencies, significantly affected the worldwide economy. It prompted an oil shock, set apart by high expansion, downturn, and energy emergencies in numerous nations.

The oil emergency of 1973 on a very basic level changed the elements of the worldwide oil market. OPEC's power as a cost setting cartel became clear, and oil-consuming countries had to defy their weakness to supply interruptions and cost instability. OPEC part nations saw a critical expansion in their oil incomes, changing their economies and giving them freshly discovered international influence.

The Second Oil Emergency (1979):

The second oil emergency, which started in 1979, was driven by various variables. A transformation in Iran prompted the fall of the Shah and the ascent of an Islamic government, causing a disturbance in Iranian oil creation. Simultaneously, the Iranian Transformation set off a flood in oil costs.

Notwithstanding the Iranian Upset, the flare-up of the Iran-Iraq Battle in 1980 further stressed worldwide oil supplies. The conflict added to a drawn out time of high oil costs, energy instability, and monetary vulnerability.

The 1979 oil emergency lastingly affected energy arrangements and security methodologies. Countries started to focus on energy preservation, effectiveness, and expansion to diminish their reliance on oil. The emergency highlighted the weakness of oil-consuming countries to international occasions in oil-creating nations.

The Tradition of the 1970s Oil Emergencies:

The oil emergencies of the 1970s had significant and getting through ramifications for the worldwide energy scene. These occasions reshaped energy strategies and prompted drives pointed toward decreasing oil reliance, for example, advancing energy proficiency, creating elective energy sources, and expanding oil holds. The emergencies likewise catalyzed endeavors to improve energy security and collaboration among countries.

OPEC's Ascent to Unmistakable quality:

The oil emergencies raised OPEC to a place of critical impact and significance in the worldwide energy market. OPEC's individuals had the option to amplify their oil incomes and influence their command over estimating and creation levels. OPEC's progress in planning its part nations and overseeing oil supplies reinforced its job as a strong cost setting cartel.

OPEC and Non-OPEC Cooperation:

Notwithstanding OPEC's activities, non-OPEC oil-creating countries assumed a vital part in answering the oil emergencies. Specifically, Mexico, Norway, and the Unified Realm got endeavors together with OPEC to settle oil costs and guarantee a predictable stock of oil to the worldwide market.

OPEC's Effect on Worldwide Energy Strategies:

OPEC's capacity to impact oil costs and creation levels significantly affected energy approaches in oil-consuming countries. Energy protection, improvement of elective energy sources, and the production of key oil holds became needs in light of OPEC's impact.

The Job of OPEC in Molding International affairs:

OPEC's ascent to noticeable quality likewise impacted the international affairs of energy. Oil-creating and consuming countries needed to adjust to another world wherein the overall influence was moving. The financial and political impact of OPEC individuals developed, and they became significant players in worldwide relations.

OPEC in the Late twentieth 100 years and Then some

The 1980s and 1990s denoted a time of change for OPEC and the worldwide oil market. Oil costs balanced out after the disturbance of the 1970s, and OPEC started to confront difficulties, like overproduction and undermining creation quantities by part nations. Furthermore, the breakdown of the Soviet Association and the finish of the Virus War had international ramifications for OPEC and the energy business.

In the late twentieth hundred years, OPEC confronted expanded contest from non-OPEC oil makers, prominently Russia, which arose as a central part in the worldwide energy market. The ascent of shale oil creation in the US likewise disturbed conventional market elements, further testing OPEC's effect on oil costs.

OPEC's impact on worldwide oil costs and supply levels stayed a subject of continuous discussion and investigation. The association's capacity to organize its individuals and oversee creation levels kept on being a basic figure the dependability of oil markets.

The 21st Hundred years and OPEC's Job Today

In the 21st hundred years, OPEC has kept on being a central member in the worldwide energy scene. The association faces developing difficulties and open doors as it explores a complex and quickly impacting universe of energy.

OPEC's job in balancing out oil advertises and affecting costs stays crucial, particularly in the midst of supply disturbances or international pressures. The association has adjusted to changing business sector elements and, on occasion, participated with non-OPEC oil makers to address worldwide energy challenges.

The journey for energy security, natural manageability, and the change to sustainable power sources have put new expectations on OPEC and its part nations. The association has perceived the need to adjust the interests of its individuals with worldwide energy and ecological objectives.

As of late, OPEC+ has arisen as a collusion among OPEC and non-OPEC oil-creating countries, including Russia. This union plays had a critical impact in overseeing worldwide oil creation and costs, featuring the continuous development of energy collaboration.

Challenges Ahead:

What's to come presents the two difficulties and open doors for OPEC. As the world movements towards practical energy sources and diminished fossil fuel byproducts, OPEC should adjust to changing interest and assumptions. The association faces the test of expanding its economies and energy portfolios to diminish their reliance on oil incomes.

OPEC likewise faces the need to oversee supply levels and estimating such that upholds the two its part nations' financial advantages and worldwide energy security. Adjusting these contending intrigues in a quickly changing energy scene is a perplexing errand.

Valuable open doors Ahead:

As the world advances to cleaner and more manageable energy sources, OPEC part nations have the potential chance to put resources into sustainable power, upgrade energy proficiency, and expand their economies. These means can lessen their weakness to variances in oil costs and request.

OPEC's part in molding worldwide energy arrangements and market solidness stays crucial. The association can add to the steadiness of oil markets and backing worldwide energy security through collaboration with non-OPEC oil makers.

5.3 The influence of oil on international relations.

Oil has been a significant and getting through force in molding worldwide relations all through the twentieth and 21st hundreds of years. The significance of oil as an essential ware couldn't possibly be more significant, as it supports financial success, innovative progressions, and military capacities of countries all over the planet. This account investigates what oil has meant for worldwide relations, looking at its part in clashes, discretion, collusions, and power elements, and how its impact has developed with changing worldwide energy patterns and needs.

The International Meaning of Oil

Oil, frequently alluded to as "dark gold," is one of the world's most significant and sought-after assets. It is a wellspring of energy, powering transportation, industry, and farming, and a urgent part in the development of plastics, synthetics, and endless buyer merchandise. The overflow and openness of oil holds have characterized the worldwide power design, and admittance to these stores has been a main thrust behind global relations and strategy.

Key Parts of Oil's Impact on Worldwide Relations:

Energy Security: The accessibility of a solid and stable stock of oil is a foundation of energy security for countries. Guaranteeing a reliable stock of oil is fundamental for financial steadiness, transportation, and public guard.

Financial Flourishing: Oil is a driver of monetary development and thriving. Countries with plentiful oil saves frequently areas of strength for have, while those vigorously dependent on oil imports might confront monetary weaknesses.

International Collusions: Oil-rich countries have frequently utilized their energy assets to produce coalitions with different nations. Significant powers have tried to tie down admittance to oil-creating locales, shaping discretionary and military collusions to safeguard their inclinations.

Military Abilities: Oil is fundamental for present day military activities. Countries with admittance to solid wellsprings of oil can support their tactical powers, while those without admittance to oil might confront impediments in their tactical abilities.

The Job of Oil in Verifiable Contentions

Oil has been a focal calculate various verifiable struggles, molding the results of wars and impacting the techniques of countries. A few striking models include:

The Second Great War and The Second Great War: Oil assumed a basic part in both universal conflicts. Command over oil fields in the Center East and different locales was an essential target for the significant powers. The English Imperial Naval force, for example, changed from coal to oil for of acquiring an upper hand. In The Second Great War, the Pivot powers, especially Germany and Japan, were moved by their restricted admittance to oil assets, which compelled their tactical tasks.

The Center East: The Center East has for some time been a point of convergence of worldwide clash because of its huge oil holds. The control of oil fields in this locale has been a driving element in territorial and global contentions, including the Bay Conflict and the Iran-Iraq War.

The Virus War: The battle for control of oil assets was a part of the more extensive Virus Battle between the US and the Soviet Association. The two superpowers tried to tie down admittance to oil-rich locales and lay out unions with oil-delivering countries.

The Bay Conflict (1990-1991): The Iraqi attack of Kuwait in 1990, roused to some extent by a craving to control Kuwait's oil holds, set off the Bay Conflict. This contention significantly affected global relations, as an alliance of countries, drove by the US, interceded to repulse the Iraqi intrusion.

Oil and Tact

The significance of oil has impacted political relations and discussions among countries. Strategic endeavors have frequently centered around tying down admittance to oil assets, guaranteeing stable supplies, and overseeing oil costs. Key discretionary exercises connected with oil include:

Respective Arrangements: Countries have gone into reciprocal arrangements to tie down admittance to oil assets. These arrangements might incorporate economic alliance, safeguard settlements, and security affirmations.

Multilateral Associations: Worldwide associations, like the Global Energy Organization (IEA) and the Association of the Petrol Sending out Nations (OPEC), play played focal parts in planning discretionary endeavors connected with oil. The IEA was made to guarantee a dependable stockpile of oil to industrialized countries, while OPEC has overseen creation levels and impacted oil costs.

Endorses and Bans: Conciliatory apparatuses like authorizes and bans have been utilized to compel oil-delivering countries to change their strategies. Authorizations might be forced to address common freedoms infringement, atomic expansion, or provincial struggles.

Intervention and Compromise: Political endeavors have looked to intercede clashes in oil-creating districts. Worldwide intercession and compromise processes mean to keep up with security and safeguard admittance to oil assets.

Energy Tact: Countries have taken part in energy discretion to reinforce energy security and expand energy sources. This incorporates arranging energy arrangements, building energy foundation, and putting resources into sustainable power projects.

Oil and Worldwide Collusions

Oil has been a basic figure the development of worldwide collusions and key organizations. Countries have looked to fall in line with oil-rich nations to guarantee admittance to oil assets, support their economies, and improve their public safety. Key instances of collusions and organizations formed by oil interests include:

The US and Saudi Arabia: The US has kept a longstanding partnership with Saudi Arabia, one of the world's biggest oil makers. This partnership has been founded on shared monetary and key interests, with the U.S. giving security confirmations to Saudi Arabia in return for admittance to its oil saves.

China and Russia: China and Russia have fostered an essential organization that incorporates energy participation. China's energy needs have driven its revenue in Russian oil and petroleum gas, prompting huge energy arrangements and speculations.

OPEC and Non-OPEC Participation: OPEC part nations have collaborated with non-OPEC oil-creating countries to oversee oil creation levels and settle costs. The OPEC+ coalition, which incorporates Russia, is an illustration of this helpful methodology.

European Association Energy Organizations: The European Association (EU) has sought after energy organizations and broadening procedures to diminish its reliance on single wellsprings of oil and petroleum gas. The EU has shaped unions with energy-rich countries and tried to construct energy foundation to upgrade energy security.

The Impact of Oil in Contemporary Global Relations

In the 21st hundred years, the impact of oil on global relations keeps on being critical, however the scene is developing. Key turns of events and contemplations include:

Energy Change: The worldwide progress to cleaner and more maintainable energy sources is reshaping global relations. Countries are looking to decrease their fossil fuel byproducts and shift away from petroleum derivatives, including oil.

Energy Broadening: Nations are differentiating their energy sources to improve energy security and lessen their reliance on oil. This remembers speculations for environmentally friendly power, atomic power, and gaseous petrol.

International Moves: The ascent of new energy players, like China, and the changing energy elements are prompting shifts in worldwide power and impact. The general significance of oil-creating countries is developing.

Natural Discretion: Ecological worries and environmental change are turning out to be progressively significant in global relations. Countries are arranging arrangements and responsibilities to address environment objectives and diminish the ecological effect of energy creation and utilization.

Energy Autonomy: A few countries are seeking after energy freedom by growing their homegrown energy creation, for example, the US with shale oil and gaseous petrol.

International Areas of interest: Oil-creating locales, like the Center East and the South China Ocean, stay international areas of interest. Arguments about oil assets and regional cases keep on affecting global relations.

Chapter 6

Environmental Challenges

Natural difficulties have become progressively squeezing in late many years, and they are probably going to linger considerably bigger in the years to come. The world is confronting a mind boggling trap of interrelated natural issues that undermine the wellbeing and prosperity of both current and people in the future. These difficulties length a large number of regions, from environmental change and biodiversity misfortune to contamination and asset consumption. In this article, we will investigate probably the most basic natural difficulties confronting our planet today and talk about the potential arrangements that can assist with resolving these issues.

One of the most pressing and irrefutable natural difficulties is environmental change. The World's environment is going through fast and extraordinary changes, fundamentally determined by human exercises like the consuming of petroleum derivatives, deforestation, and modern cycles. These exercises discharge ozone depleting substances, including carbon dioxide and methane, into the air, which trap intensity and lead to a progressive expansion in worldwide temperatures. This peculiarity, known as an unnatural weather change, has extensive ramifications for the planet.

As temperatures increase, we are seeing a scope of unfriendly impacts. Liquefying ice covers and glacial masses add to rising ocean levels, which can immerse beach front networks and low-lying islands. More incessant and extreme climate occasions, like storms, dry spells, and heatwaves, are turning into the new standard, causing obliterating influences on networks, biological systems, and economies. The disturbance of environment examples can likewise prompt rural difficulties, influencing food creation and food security for billions of individuals.

Notwithstanding environmental change, biodiversity misfortune is another basic natural test. Human exercises, like living space obliteration, overexploitation of normal assets, and the presentation of intrusive species, are driving the termination of plant and creature species at a disturbing rate. This deficiency of biodiversity has significant ramifications for environments and the administrations they give to human social orders.

Biodiversity is fundamental for the security and strength of biological systems. It adds to the sanitization of air and water, the fertilization of yields, and the guideline of environment. Besides, it offers likely well-springs of new medications, materials, and hereditary variety that can assist with tending to a scope of human requirements. At the point when biodiversity is undermined, these fundamental administrations and assets are compromised, and the drawn out prosperity of mankind is jeopardized.

Contamination is one more major natural test. Contamination takes many structures, including air contamination, water contamination, and soil pollution. Modern cycles, transportation, and horticulture discharge various poisons into the climate, with ramifications for both human wellbeing and biological systems. For instance, air contamination can prompt respiratory illnesses, while water contamination can sully drinking water sources and mischief oceanic life.

The aggregation of plastic waste in our seas has acquired critical consideration lately. Single-use plastics, for example, sacks and containers, persevere in the climate for quite a long time, separating into more modest particles that are ingested by marine creatures and enter the well established pecking order. This contamination hurts marine life as well as stances dangers to human wellbeing through the utilization of tainted fish.

Asset exhaustion is intently attached to the issue of contamination. Human utilization designs, particularly in created nations, have prompted the overexploitation of normal assets, like freshwater, woods, and minerals. This abuse is unreasonable, as it drains limited assets and upsets the equilibrium of environments. For instance, deforestation adds to territory misfortune and diminishes the World's ability to retain carbon dioxide, worsening environmental change.

Water shortage is another squeezing ecological test. While the Earth is shrouded in water, just a little level of it is freshwater, and quite a bit of that is secured in glacial masses and distant. As the worldwide populace keeps on developing, the interest for freshwater for drinking, agribusiness, and industry increments. Environmental change further worsens water shortage by adjusting precipitation designs and expanding the gamble of dry season in certain areas.

The inconsistent appropriation of water assets additionally presents difficulties. In many regions of the planet, water shortage is a reality, prompting contest for restricted water supplies and possible contentions. Guaranteeing admittance to spotless and safe water for everything is a worldwide test that requires compelling water the board and preservation endeavors.

The deficiency of arable land and soil corruption is a developing worry too. Soil is a limited asset that is fundamental for food creation. Concentrated farming practices, like over the top utilization of substance composts and pesticides, can exhaust soil supplements and damage its design. Disintegration, brought about by deforestation and unfortunate land the board, can prompt the deficiency of ripe dirt. Soil debasement compromises the drawn out supportability of agribusiness and food security.

The issue of waste administration is firmly connected with asset consumption and contamination. The direct model of take, make, and arrange has prompted a developing measure of waste created by human social orders. Landfills are topping off, and cremation of waste can deliver destructive contaminations very high. A shift towards a round economy, where assets are reused and reused, is significant for diminishing waste and limiting the ecological effect of our utilization designs.

The urbanization of the total populace is another ecological test. As additional individuals move to urban communities looking for better open doors, metropolitan regions extend, consuming important land and assets. The convergence of individuals in metropolitan habitats can prompt issues connected with congestion, air contamination, and the interest for foundation and administrations. Metropolitan preparation and maintainable city advancement are fundamental to relieve the natural effect of urbanization.

The convergence of natural difficulties with social and financial issues is complicated. Ecological debasement frequently lopsidedly influences helpless and minimized networks. Low-pay populaces might endure the worst part of contamination, experience food and water uncertainty, and need admittance to medical care and catastrophe readiness assets. Tending to natural difficulties requires a simply change that thinks about the necessities and privileges, everything being equal, particularly those generally defenseless against ecological effects.

The interconnected idea of these natural difficulties requires all encompassing arrangements. One of the vital procedures for resolving these issues is a shift towards maintainable turn of events. Feasible improvement looks to adjust financial development, social prosperity, and natural insurance. It advances rehearses that address the issues of the present without compromising the capacity of people in the future to address their own issues.

Environmentally friendly power sources assume a critical part in relieving environmental change. Changing from petroleum products to clean energy sources, for example, sun based, wind, and hydroelectric power, can diminish ozone depleting substance outflows and lessening the dependence on limited assets. The turn of events and reception of energy-proficient innovations can additionally upgrade the supportability of energy use.

Preservation and reclamation of regular environments are fundamental for biodiversity assurance. Safeguarded regions, like public parks and marine stores, give environments to assorted species and assist with keeping up with hereditary variety.

Once again introducing species into their local territories and carrying out protection projects can support species recuperation. Reasonable land use works on, including agroforestry and natural cultivating, can decrease environment obliteration and advance biological system wellbeing.

Tending to contamination requires administrative measures and innovative headways. State run administrations and global associations can lay out ecological regulations and guidelines to restrict poison discharges and guarantee mindful waste administration. Mechanical developments, for example, clean transportation innovations and waste-to-energy processes, can likewise add to contamination decrease.

Asset exhaustion can be relieved through asset productivity and round economy rehearses. Decreasing asset squander through reusing and reusing materials can broaden the existence of limited assets. Practical ranger service and fisheries the board can guarantee the dependable utilization of woods and seas. Moreover, customer decisions and conduct assume a vital part in decreasing asset utilization.

Water shortage can be tended to through superior water the executives and preservation measures. Interests in water foundation, like water treatment and dissemination frameworks, can assist with giving admittance to perfect and safe water. Horticulture rehearses that utilization water all the more proficiently, like trickle water system, can diminish water squander. Moreover, securing and reestablishing watersheds can improve water quality and accessibility.

6.1 The environmental impact of oil extraction.

The natural effect of oil extraction is a subject of significant concern and examination as the world's reliance on petroleum derivatives proceeds. Oil is an important and flexible asset, yet its extraction and creation come at a tremendous expense for the climate. In this article, we will dive into the different natural outcomes related with oil extraction, from the underlying boring to the transportation and utilization of oil. It is fundamental to perceive the extent of these effects and consider elective wellsprings of energy and feasible practices to relieve the harm.

The main period of oil extraction, boring, is joined by a scope of ecological difficulties. The penetrating system includes making boreholes in the world's surface to get to underground repositories of oil. This action can prompt living space disturbance, particularly in perfect or delicate conditions. While penetrating happens in environmentally significant districts, like rainforests, wetlands, or Cold biological systems, it can devastatingly affect neighborhood untamed life and plant species. These environments frequently have one of a kind and fragile animal types that may not endure the unsettling influences achieved by penetrating exercises.

Moreover, boring tasks normally require the development of streets, penetrating cushions, and other foundation. These improvements part living spaces and can prompt the confinement of untamed life populaces.

For instance, in the Amazon rainforest, oil extraction has been connected to deforestation and the annihilation of basic territories, imperiling species like pumas, ungulates, and different bird species.

Notwithstanding territory disturbance, the penetrating system can likewise represent a gamble of oil slicks. Inadvertent spills during boring, transportation, or capacity can have disastrous ramifications for the climate. Oil is exceptionally poisonous to sea-going life, and spills can prompt mass fish kills, harm to coral reefs, and long haul mischief to oceanic biological systems. The Exxon Valdez oil slick in 1989 and the Deepwater Skyline oil slick in 2010 act as unmistakable tokens of the natural demolition brought about by oil slicks.

One more critical ecological issue connected with boring is the infusion of wastewater into removal wells. In pressure driven cracking, or deep earth drilling, a procedure used to remove oil and gas from shale developments, huge volumes of water blended in with synthetics are infused into the ground to fall to pieces shakes and delivery hydrocarbons. This cycle produces significant measures of wastewater, which are frequently discarded by infusing them into profound underground wells. The infusion of wastewater into removal wells has been connected to prompted seismicity or man-made tremors, which can harm framework and posture dangers to public security.

The extraction of oil likewise prompts the arrival of methane, a strong ozone depleting substance. Methane is many times delivered during the boring system and from spillages in hardware and framework. While carbon dioxide (CO2) is the most notable ozone depleting substance, methane has a lot more noteworthy transient warming impact. It adds to environmental change by catching intensity in the climate, fueling the worldwide temperature increment. The ecological results of methane discharges are significant, and they highlight the need to address these holes and diminish methane discharge from oil extraction tasks.

Transportation of oil is one more stage in the oil creation process with ecological ramifications. Oil is commonly moved through pipelines, big haulers, and trucks, every one of which presents interesting natural dangers. Pipelines, while thought about a generally proficient and practical method of transport, can prompt oil slicks on the off chance that they break or hole. These spills can have unfortunate results, as exhibited by the Dakota Access Pipeline fights and the Cornerstone XL Pipeline contention in North America. Pipelines likewise require land clearing, possibly influencing neighborhood environments and living spaces.

Oil big haulers are one more typical technique for moving oil across the world's seas. Mishaps including oil big haulers, like groundings or crashes, can prompt huge oil slicks. The Exxon Valdez calamity, referenced prior, involved a grounded oil big hauler that spilled great many gallons of unrefined petroleum into the Sovereign William Sound, causing serious natural harm. The gamble of oil slicks from big haulers stays a huge worry for marine conditions.

Truck transportation of oil presents its own arrangement of difficulties. It includes the utilization of street organizations and can prompt mishaps and spills ashore. Also, the burning of diesel fuel in trucks adds to air contamination, delivering particulate matter, nitrogen oxides, and unstable natural mixtures into the climate, which can antagonistically affect human wellbeing and neighborhood air quality.

The utilization of oil for energy creation and as a fuel source has broad and significant natural effects. The ignition of oil in power plants and vehicles radiates a scope of toxins up high. These toxins incorporate sulfur dioxide (SO_2), nitrogen oxides (NOx), particulate matter (PM), and unstable natural mixtures (VOCs). These emanations add to air contamination and affect human wellbeing, including respiratory issues, cardiovascular infections, and, surprisingly, unexpected passing.

Moreover, the arrival of carbon dioxide (CO_2) from consuming oil is an essential driver of environmental change. As more oil is consumed for energy, the centralization of CO_2 in the air increments, bringing about a continuous climb in worldwide temperatures. The results of environmental change incorporate more successive and serious heatwaves, outrageous climate occasions, ocean level ascent, and disturbances to biological systems. These progressions can have flowing consequences for biodiversity, agribusiness, and the accessibility of freshwater assets.

The natural effect of oil extraction reaches out to the social and social elements of networks living close to extraction locales. Native people group, specifically, frequently endure the worst part of oil advancement. Their conventional terrains might be infringed after, prompting the deficiency of social legacy and profound importance. The interruption of neighborhood environments can likewise influence conventional means rehearses and undermine the livelihoods of native individuals who depend on the land for their endurance.

Oil extraction can bring about friendly contentions, especially in areas with restricted government oversight and powerless guidelines. Clashes might emerge because of disagreements about land proprietorship, admittance to assets, and the dispersion of monetary advantages from oil extraction. Such struggles can have serious outcomes, prompting savagery and shakiness in impacted locales.

The natural effects of oil extraction stretch out past the creation stage and go on all through the lifecycle of oil. For example, the refining of unrefined petroleum into oil based goods creates extra ecological difficulties. The refining system includes the utilization of energy-concentrated processes and the arrival of air contaminations. It additionally delivers perilous waste, for example, spent impetuses and slops, which require legitimate removal and the board to forestall tainting of soil and water.

The end-utilization of oil based goods likewise adds to natural difficulties. Vehicles controlled by gas and diesel energizes discharge ozone harming substances and air poisons while working.

The auto business is progressively changing to electric vehicles (EVs) to diminish discharges and dependence on oil. EVs offer an all the more harmless to the ecosystem elective, as they produce no tailpipe discharges and can be accused of power from cleaner and inexhaustible sources.

Endeavors to relieve the ecological effect of oil extraction are basic. Changing to cleaner and sustainable power sources is a principal procedure for diminishing the world's reliance on oil. Sunlight based, wind, and hydroelectric power, among others, offer practical and low-outflow options for producing power. Putting resources into innovative work of cutting edge energy capacity advances and expanding energy proficiency can additionally lessen the requirement for oil in the energy area.

Stricter guidelines and oversight are fundamental to guarantee that oil extraction and transportation tasks satisfy ecological guidelines.

Carrying out accepted procedures, for example, spill discovery and fix programs, can assist with decreasing methane emanations from oil and gas activities. Further developed spill reaction and cleanup measures can limit the natural harm brought about by oil slicks.

Integrating ecological contemplations into land-use arranging and effect appraisals is urgent. Distinguishing and safeguarding biologically delicate regions can assist with saving significant living spaces and species. Also, capable land recovery and environment reclamation practices ought to be a vital part.

6.2 Oil spills and their consequences.

Oil slicks are ecological fiascos that have broad ramifications for biological systems, natural life, and human populaces. These episodes happen when oil is delivered into the climate, regularly through unintentional spills from boats, pipelines, or oil boring stages. The effect of an oil slick is quick and frequently pulverizing, presenting huge difficulties for cleanup and long haul recuperation. In this exposition, we will investigate the causes, results, and reactions to oil slicks, revealing insight into the pressing requirement for avoidance and alleviation.

Oil slicks can be set off by different elements, yet they most normally happen because of mishaps during the transportation of oil, penetrating and extraction tasks, or big hauler and pipeline bursts. The Deepwater Skyline oil slick in the Bay of Mexico in 2010 is a conspicuous illustration of a devastating oil slick brought about by a penetrating stage blast. This occurrence delivered huge number of barrels of oil into the sea, bringing about broad harm to marine biological systems and beach front areas.

Boats and big haulers moving oil across the world's seas are especially vulnerable to oil slicks. Mishaps like impacts, groundings, or gear disappointments can prompt oil being delivered into the water. The Exxon Valdez oil slick in 1989, in which an oil big hauler steered into the rocks in Gold country, stays one of the most scandalous oil slicks ever.

It brought about the arrival of almost 11 million gallons of unrefined petroleum into the unblemished waters of Ruler William Sound, causing extreme natural and monetary harm.

Oil pipelines, while thought about a more proficient method of transport, are not invulnerable to cracks or holes. The Cornerstone Pipeline spill in North Dakota in 2019 fills in as a new illustration of a pipeline release, delivering in excess of 380,000 gallons of raw petroleum into the climate. These occurrences feature the gamble of oil transportation through pipelines, particularly when they navigate naturally touchy or thickly populated regions.

When an oil slick happens, its quick and long haul outcomes can be broad. The essential effect is on the marine climate. Oil slicks bring about the arrival of poisonous hydrocarbons into the water, which can devastatingly affect marine life. Oil-covered birds, ocean turtles, and marine well evolved creatures frequently become banner creatures for the noticeable damage brought about by oil slicks. These creatures can become covered in oil, lose lightness, and experience the ill effects of hypothermia as their normal protection is compromised. Ingesting oil can prompt inner harm, harming, and demise.

Oil slicks additionally hurt fish and other oceanic life. Spilled oil can choke out fish eggs and hatchlings, keeping them from incubating or getting by. The poisons in oil can collect in fish and shellfish, entering the pecking order and presenting wellbeing dangers to people who eat sullied fish. Besides, oil slicks can harm coral reefs, significant reproducing and taking care of justification for the overwhelming majority marine species, prompting long haul influences on biodiversity.

Seaside environments are especially powerless against the outcomes of oil slicks. Mangroves, estuaries, and wetlands are basic natural surroundings for various species and act as nurseries for fish and shellfish. At the point when oil enters these regions, it can cover vegetation, upset environments, and damage the organic entities that depend on these territories. The drawn out recuperation of seaside environments can require many years, and a few regions might in all likelihood never completely recuperate.

Past the quick natural outcomes, oil slicks additionally have huge financial ramifications. Seaside people group frequently depend on fishing and the travel industry, and when these businesses are upset by oil tainting, jobs are risked. The expense of cleanup endeavors and the rebuilding of impacted regions can falter. For instance, the Deepwater Skyline oil slick cleanup cost billions of dollars, and the financial effect on the Bay Coast was felt for a really long time.

Oil slicks can likewise have significant social and social outcomes, especially for native networks. Numerous native populaces rely upon customary resource rehearses, like hunting, fishing, and assembling, for their food and social personality. At the point when oil slicks taint their customary terrains and waters, these networks face monetary misfortunes as well as the disintegration of their social legacy.

Notwithstanding the natural and monetary repercussions, oil slicks raise serious general wellbeing concerns. Openness to oil and the synthetic compounds utilized in cleanup endeavors can antagonistically affect human wellbeing. Those engaged with cleanup activities are in danger of skin and respiratory issues, as well as openness to poisonous synthetic substances. Networks close to oil slick locales might confront long haul wellbeing chances, as well as interruptions to their day to day routines and admittance to clean water and food.

The natural effect of oil slicks is enduring, and recuperation can be slow and testing. Successful reaction and cleanup endeavors are fundamental to limit the harm. These endeavors regularly include the utilization of different strategies and advances, including the sending of oil blasts to contain and gather surface oil, the utilization of skimmers to eliminate oil from the water, and the use of dispersants to separate oil spills. At times, controlled consumes are directed to eliminate surface oil.

While cleanup is basic, it isn't generally imaginable to eliminate all the spilled oil, and some of it might continue in the climate for quite a long time. This excess oil can keep on delivering poisons and have progressing biological effects. Normal cycles, like microbial corruption, can assist with separating oil after some time, however this can be a sluggish and inadequate interaction.

Notwithstanding mechanical and synthetic cleanup strategies, there is developing interest in the utilization of bioremediation procedures. Bioremediation includes the utilization of microorganisms, like microbes and organisms, to separate oil and its parts. These microorganisms can normally happen in the climate or can be acquainted with speed up the debasement cycle. While bioremediation shows guarantee, its viability can be impacted by factors like temperature, oxygen levels, and the presence of different pollutants.

Avoidance is the best way to deal with alleviating the natural effect of oil slicks. Thorough wellbeing measures, normal assessments, and upkeep of foundation and gear are crucial for diminishing the gamble of spills during oil penetrating, transportation, and capacity. Propels in innovation, like twofold hulled big haulers and further developed spill reaction hardware, have helped upgrade the wellbeing of oil transportation.

Administrative oversight and requirement are vital to guaranteeing that oil industry rehearses fulfill natural guidelines. State run administrations and worldwide associations lay out rules and guidelines to forestall and answer oil slicks. These guidelines cover regions, for example, vessel plan and wellbeing, spill reaction arranging, and risk for harms. Adherence to these guidelines can altogether decrease the gamble of oil slicks and cutoff their results.

As of late, there has been a developing push for the turn of events and utilization of option and sustainable power sources. Changing away from petroleum products, including oil, is a basic move toward lessening the natural dangers related with oil extraction and transportation.

Sun oriented, wind, and hydroelectric power, among others, offer more reasonable and harmless to the ecosystem wellsprings of energy.

As the world changes to cleaner energy choices, the interest for oil can be diminished, decreasing the gamble of oil slicks.

The outcomes of oil slicks are huge, enduring, and frequently terrible. They help us to remember the requirement for a complete way to deal with counteraction, readiness, and reaction. Interests in examination and innovation can prompt superior spill reaction abilities, including more compelling control and cleanup strategies. Moreover, worldwide collaboration and shared liability in tending to the dangers of oil slicks are vital, as oil is a worldwide product, and its effect knows no boundaries.

Besides, the utilization of financial instruments, like expenses and charges on oil creation and transportation, can make impetuses for organizations to put resources into wellbeing and avoidance measures. These assets can likewise be reserved for spill reaction and rebuilding endeavors, guaranteeing that those answerable for spills bear the monetary weight of their outcomes.

6.3 The push for environmental sustainability.

The push for natural supportability has picked up speed lately as social orders overall perceive the earnestness of tending to ecological difficulties, monitoring assets, and relieving the effects of environmental change. Natural supportability envelops an expansive scope of practices, strategies, and ways of behaving pointed toward protecting the planet's biological systems, decreasing fossil fuel byproducts, and guaranteeing a livable and sound world for current and people in the future. In this paper, we will investigate the vital drivers behind the push for ecological manageability, the rules that support it, and the basic advances expected to accomplish a more practical future.

One of the essential drivers behind the push for natural manageability is the acknowledgment of the interconnectedness of biological frameworks and human prosperity. The climate is certainly not a different element from society; it is inherently connected to human wellbeing, flourishing, and generally speaking personal satisfaction. Corruption of the climate, whether through deforestation, air and water contamination, or the consumption of regular assets, has extensive ramifications for biological systems and, at last, human social orders.

Environmental change is a critical and squeezing worry that embodies this interconnectedness. The consuming of non-renewable energy sources, deforestation, and modern cycles discharge ozone depleting substances into the air, prompting an unnatural weather change and the disturbance of environment designs. Climbing temperatures, outrageous climate occasions, and ocean level ascent present direct dangers to human networks, horticulture, and economies. Perceiving the reliance of ecological and cultural prosperity is a crucial driver of the push for manageability.

Natural manageability is established in a few key rules that guide its execution. One of the focal standards is the idea of stewardship, which underscores dependable administration and care of the World's normal assets. This approach requires the mindful utilization of assets, the assurance of biodiversity, and the decrease of waste and contamination. Stewardship recognizes that humankind plays a part in protecting the climate for people in the future.

Another center guideline is the preparatory rule, which inclinations proactive measures to forestall mischief to the climate and human wellbeing. It highlights the significance of making a move when the potential for hurt is conceivable, regardless of whether logical sureness isn't yet settled. The preparatory rule has been instrumental in driving guidelines and arrangements pointed toward decreasing the natural effect of different exercises, like synthetic substances, hereditarily changed creatures, and modern cycles.

Value and civil rights are likewise crucial standards of natural supportability. Perceiving that ecological debasement frequently excessively influences minimized and weak networks, manageability endeavors look to address these differences. The idea of natural equity accentuates the fair dispersion of ecological advantages and weights, guaranteeing that all networks approach clean air, water, and solid conditions.

Also, supportability standards underscore the significance of inter-generational value. This standard calls for present ages to consider the prosperity of people in the future in their navigation. It highlights the moral obligation to leave a planet that can uphold the requirements and desires of the people who will acquire it. This point of view is pivotal for long haul arranging and the preservation of assets.

The push for natural supportability likewise includes a few fundamental methodologies and activities that can assist with relieving ecological difficulties and advance a more maintainable future. A portion of these procedures include:

Progressing to Sustainable power: One of the main strides toward natural manageability is moving from petroleum derivatives to sustainable power sources, for example, sun oriented, wind, and hydropower. Environmentally friendly power advances offer cleaner and more feasible other options, decreasing ozone harming substance discharges and reliance on limited assets.

Energy Productivity: Further developing energy effectiveness in modern, private, and transportation areas is imperative. Energy-proficient innovations, building plans, and transportation frameworks can fundamentally decrease energy utilization and discharges.

Preservation and Reasonable Asset The executives: Safeguarding normal territories, moderating biodiversity, and it are fundamental for carry out economical asset the board rehearses. This incorporates economical ranger service and fisheries rehearses, living space reclamation, and the assurance of basic environments.

Squander Decrease and Reusing: Lessening waste and expanding reusing endeavors can limit the natural effect of asset utilization. The shift from a straight "take-make-arrange" model to a roundabout economy, where assets are reused and reused, is vital.

Reasonable Horticulture: Carrying out supportable cultivating rehearses, like natural agribusiness, agroforestry, and regenerative cultivating, can decrease the ecological effect of farming, including soil corruption and water contamination.

Reforestation and Afforestation: Establishing trees and reestablishing woods can assist with sequestering carbon dioxide, further develop air and water quality, and give natural surroundings to untamed life.

Clean Transportation: Advancing public transportation, electric vehicles, and trekking and strolling framework can diminish discharges and further develop air quality in metropolitan regions.

Feasible Metropolitan Preparation: Planning urban areas and metropolitan regions considering supportability, including green spaces, energy-effective structures, and productive transportation frameworks, can prompt more maintainable and reasonable networks.

Carbon Estimating: Executing carbon evaluating instruments, for example, carbon duties or cap-and-exchange frameworks, can boost organizations and people to lessen their fossil fuel byproducts.

Strategy and Guideline: Legislatures assume a vital part in setting and upholding ecological guidelines and arrangements that advance supportability. These strategies can incorporate outflows guidelines, preservation regulations, and impetuses for clean advancements.

Instruction and Mindfulness: Raising public mindfulness and giving training about ecological maintainability are fundamental for drive individual and aggregate activities. Understanding the outcomes of ecological corruption can inspire individuals to pursue practical decisions.

Worldwide Collaboration: Ecological difficulties frequently rise above public limits. Peaceful accords and collaboration are important to resolve worldwide issues, for example, environmental change, biodiversity misfortune, and marine contamination.

Mechanical Development: Headways in innovation can prompt more economical arrangements, from cleaner energy advances to imaginative waste administration strategies.

The push for ecological maintainability likewise includes the commitment of organizations and ventures in dependable practices. Numerous organizations are perceiving the advantages of maintainability, including cost investment funds, improved standing, and admittance to new business sectors.

Manageable strategic policies envelop ecological stewardship, social obligation, and financial reasonability. Organizations are embracing reasonable production network the executives, diminishing waste, and integrating manageability into their center business techniques.

Notwithstanding individual and corporate endeavors, government activity is pivotal for accomplishing natural maintainability. State run administrations have the ability to order regulations, guidelines, and arrangements that drive supportability at the public and worldwide levels. They can likewise give impetuses and backing to environmentally friendly power, energy proficiency, and protection endeavors.

Worldwide collaboration is essential for tending to worldwide natural difficulties. Arrangements, for example, the Paris Settlement on environmental change and the Show on Organic Variety mean to unite countries to set targets and make an aggregate move to moderate environmental change and safeguard biodiversity. These arrangements underscore the common obligation, everything being equal, to resolve ecological issues.

The push for natural maintainability isn't without its difficulties and obstructions. Obstruction from personal stakes, political polarization, and financial limitations can upset progress. Be that as it may, the criticalness of ecological difficulties, for example, environmental change, biodiversity misfortune, and asset exhaustion, requests a deliberate and resolute work to accomplish maintainability. The outcomes of inaction are too grave to even consider overlooking.

Endeavors to progress natural manageability require cooperation at all degrees of society - from people and networks to organizations, state run administrations, and global associations. Each activity, regardless of how little, can add to the aggregate work to safeguard the planet and guarantee a more economical future for all. By embracing maintainability as a core value and pursuing it a center piece of choice making, humankind can cooperate to address ecological difficulties and make an existence where nature and society flourish as one.

7

Chapter 7

The Quest for Alternatives

In a world overflowing with difficulties and vulnerabilities, the mission for options has turned into a consistently present and squeezing try. Mankind remains at an intersection, confronting issues that request imaginative arrangements. These difficulties range the domains of environmental change, energy, medical care, instruction, and then some, expecting us to investigate new skylines and look for options that can make ready for a more splendid, more maintainable future.

One of the most impressive difficulties facing humankind today is environmental change. The constant expansion in ozone depleting substance discharges has upset the fragile equilibrium of our planet's biological system, bringing about climbing worldwide temperatures, more successive outrageous climate occasions, and the liquefying of polar ice covers. The outcomes are significant, influencing the climate as well as our economies, social orders, and the prosperity of people in the future.

To battle this existential danger, a worldwide work to find options in contrast to our ongoing methods of energy creation and utilization is in progress. The progress from petroleum products to environmentally friendly power sources, for example, sun based, wind, and hydroelectric power, is an essential move toward lessening fossil fuel byproducts. Advancements in energy capacity and dissemination frameworks are similarly significant, as they can make the irregular idea of sustainable power sources more solid and effective. The journey for options in the energy area isn't just a natural goal yet additionally a financial one, as it can make new enterprises and open positions while lessening our reliance on limited assets.

Medical services is one more space where the quest for choices is vital. The Coronavirus pandemic has featured the weaknesses of our medical care frameworks, provoking a reconsideration of our ways to deal with sickness counteraction, therapy, and medical care conveyance. Telemedicine, for instance, arose as an option in contrast to conventional in-person medical care, permitting patients to remotely talk with medical care experts. Furthermore, the improvement of mRNA-based immunizations, similar to the ones used to battle the Covid, addresses a momentous option in contrast to customary immunization advances. These choices offer promising roads for further developing medical services access, proficiency, and adequacy, both in pandemic reaction and routine consideration.

In the field of schooling, the mission for options has picked up speed in light of the changing requirements of students and the advancing scene of data and innovation. Web based learning stages have flooded in prominence, offering adaptable and open options in contrast to conventional physical training. This shift has likewise started a reconsidering of teaching method, with instructors investigating elective showing strategies, for example, project-based learning, customized guidance, and experiential schooling, to more readily draw in and plan understudies for the difficulties representing things to come.

The monetary area isn't resistant to the call for options. Digital forms of money, like Bitcoin and Ethereum, have acquired unmistakable quality as elective types of advanced cash and speculation. These decentralized frameworks offer new open doors for monetary consideration and financial strengthening, testing the customary banking and installment frameworks. As monetary innovations keep on developing, they can possibly give options that democratize admittance to monetary administrations and reshape the worldwide monetary scene.

Moreover, the mission for choices reaches out into farming and food creation, as we wrestle with issues of food security, asset supportability, and moral worries. Elective protein sources, for example, plant-based and lab-developed meats, are acquiring fame as feasible options in contrast to conventional creature horticulture. These developments can possibly lessen the ecological effect of food creation, lighten creature government assistance concerns, and guarantee a more practical worldwide food supply.

Transportation is one more space in which options are direly required. Customary gas powered motor vehicles are a significant wellspring of fossil fuel byproducts, adding to air contamination and environmental change. The improvement of electric vehicles (EVs) and the development of public transportation choices, similar to high velocity rail, are significant stages in decreasing the ecological impression of transportation. In addition, the journey for choices in transportation incorporates independent vehicles, which can possibly upgrade security, decrease gridlock, and further develop versatility for those with restricted admittance to transportation.

The difficulties we face in the 21st century are not restricted to those referenced previously. Issues like urbanization, water shortage, squander the executives, and worldwide wellbeing emergencies require a nonstop and extensive investigation of options.

The intricacy of these difficulties requests that we approach them with a mentality of development, variation, and versatility. The journey for choices isn't just a choice however a need in the event that we are to fabricate a more feasible, fair, and strong future for us and people in the future.

The criticalness of finding choices is highlighted by the possible outcomes of inaction. Environmental change, whenever left neglected, could prompt more extreme climate occasions, rising ocean levels, and the removal of millions of individuals. Medical services frameworks unprepared to answer pandemics and other wellbeing emergencies might bring about pointless affliction and death toll. Schooling systems that neglect to adjust to changing requirements could leave understudies caught off guard for the requests of a quickly developing world. Monetary frameworks that oppose development might pass up new open doors for development and success. Farming and food creation that disregard supportability and moral worries might fuel worldwide difficulties like yearning and ecological corruption. Transportation frameworks that keep on depending on petroleum derivatives might demolish air quality and add to the continuous environment emergency.

The journey for options isn't without its difficulties. Protection from change, personal stakes, and asset limitations can block the turn of events and reception of elective arrangements. Moreover, the way to finding options is frequently cleared with vulnerability and hazard. In any case, history has shown that people are astoundingly skilled at development when confronted with existential difficulties. Our capacity to adjust, develop, and conquer impediments has prompted groundbreaking forward leaps from the beginning of time. The journey for options is, in numerous ways, a demonstration of the dauntless human soul and our ability for progress.

One of the essential drivers of development in the mission for options is innovation. Progresses in data innovation, man-made reasoning, biotechnology, and materials science have extended the range of potential outcomes in pretty much every field. These innovative progressions empower us to investigate new boondocks, find novel arrangements, and streamline existing cycles. For instance, the utilization of computerized reasoning in medical services can help with diagnosing sicknesses and creating customized therapy plans. In agribusiness, accuracy cultivating strategies, empowered by information examination and sensor innovation, can streamline crop yields and decrease asset use. Also, blockchain innovation offers straightforwardness and security in monetary exchanges, testing customary financial frameworks. The proceeded with advancement and use of these and different advancements hold extraordinary commitment in our quest for options.

Also, the journey for options is intently attached to the standards of supportability. Supportability involves tracking down arrangements that address recent concerns without compromising the capacity of people in the future to address their own issues. It envelops natural, monetary, and social aspects, and its standards underlie numerous elective methodologies.

For example, sustainable power sources advance ecological maintainability by lessening fossil fuel byproducts, while additionally adding to monetary supportability by making position in the perfect energy area. Additionally, supportable agribusiness rehearses expect to safeguard normal assets, safeguard biodiversity, and give evenhanded admittance to nutritious food. The quest for manageability is necessary to the mission for options in all areas.

In the quest for options, interdisciplinary joint effort is frequently pivotal. The intricate difficulties we face seldom fit perfectly inside the limits of a solitary field or discipline. Addressing these difficulties frequently requires the skill of researchers, engineers, financial experts, sociologists, policymakers, and numerous others cooperating. Interdisciplinary coordinated effort cultivates a comprehensive comprehension of the main things in need of attention and energizes imaginative critical thinking. It permits us to think about a more extensive scope of choices and their possible outcomes.

Notwithstanding interdisciplinary coordinated effort, worldwide participation is fundamental. A large number of the difficulties we face are transnational in nature. Environmental change, for instance, is a worldwide issue that requires composed endeavors from nations all over the planet. The Paris Arrangement, a worldwide deal pointed toward restricting a dangerous atmospheric devation, epitomizes the significance of worldwide collaboration in tending to such difficulties. Also, worldwide wellbeing emergencies, similar to the Coronavirus pandemic, request global joint effort in regions, for example, antibody dispersion, data sharing, and exploration.

Moreover, the mission for options is interlaced with moral contemplations. At the point when we look for options, we should think about their specialized possibility and financial suitability as well as their moral ramifications. Our decisions in quest for choices can have significant moral and social outcomes. For example, the advancement of independent vehicles brings up issues about wellbeing, security, and the likely removal of laborers in the transportation .

7.1 The search for alternative energy sources.

In a world confronting consistently expanding energy requests and developing worries over ecological supportability, the quest for elective energy sources has become the dominant focal point. As the results of depending on non-renewable energy sources become progressively clear, the basic to find cleaner, more economical options develops more earnest. This journey for elective energy sources envelops a large number of innovations and methodologies, determined to moderate environmental change, diminishing contamination, improving energy security, and encouraging financial turn of events.

The essential driver for looking for elective energy sources is the worldwide test of environmental change. The consuming of petroleum derivatives, like coal, oil, and flammable gas, for energy creation discharges ozone harming substances, especially carbon dioxide (CO_2), into the air.

These gases trap intensity and prompt worldwide temperatures to climb, bringing about an outpouring of results, including more successive and extreme climate occasions, ocean level ascent, and changes in environments. The direness of tending to environmental change is clear, and finding elective energy sources is a basic part of moderating its belongings.

One of the most generally perceived elective energy sources is sunlight based power. Sun oriented energy saddles the force of the sun's beams, which gives a plentiful, sustainable, and essentially unlimited wellspring of energy. Sun powered chargers, otherwise called photovoltaic (PV) cells, convert daylight into power. The innovation has progressed quickly lately, making sun based power progressively available and savvy. Sun oriented establishments can be tracked down on private roofs, in gigantic sun based ranches, and in metropolitan settings, giving clean energy to drive homes and organizations.

Wind energy is another noticeable elective energy source that has built up some decent momentum. Wind turbines catch the active energy of the breeze and convert it into power. Enormous scope wind ranches, both coastal and seaward, have been created in districts with steady wind designs. Wind power enjoys the benefit of being harmless to the ecosystem and sustainable, yet it is dependent upon varieties in wind speed and heading. Progresses in turbine innovation and network coordination have assisted make with winding energy a more dependable wellspring of force.

Hydropower, otherwise called hydroelectric power, uses the energy of streaming water to create power. Dams and supplies are developed to control the progression of water, which is guided through turbines to deliver electrical power. Hydropower is a deeply grounded wellspring of elective energy and has been a critical supporter of worldwide power age for a long time. It is dependable, delivers no immediate ozone depleting substance discharges, and can be changed in accordance with satisfy shifting energy needs. Be that as it may, its natural effects, like the interruption of oceanic environments and dislodging of networks, have incited a quest for additional maintainable ways to deal with hydropower.

Geothermal energy takes advantage of the World's interior intensity by utilizing heat siphons to move warmth from the beginning structures or by outfitting the steam and high temp water found underneath the World's surface to create power. Geothermal power plants are normally situated in districts with huge geothermal action, like fountains and underground aquifers. This elective energy source is profoundly dependable, accessible every minute of every day, and creates exceptionally low ozone harming substance emanations. As innovation propels, the potential for saddling geothermal energy extends to areas past those with clear geothermal elements.

Biomass energy gets from natural materials, like wood, agrarian deposits, and even green growth. Biomass can be singed to create heat or changed over into biofuels, as biodiesel or ethanol, to drive vehicles.

While biomass is considered inexhaustible on the grounds that the plants used to create it very well may be regrown, it can in any case add to natural issues, like deforestation and rivalry for land and assets. Trend setting innovations are arising to moderate these difficulties and make biomass energy more supportable.

Another elective energy source not too far off is flowing and wave energy. Tides and waves, driven by the gravitational powers of the moon and the sun, offer a steady and unsurprising wellspring of energy. Flowing energy is tackled by catching the development of water during elevated and low tides, while wave energy is caught by gadgets that move with the movement of the waves. These advances are still in the beginning phases of improvement yet hold guarantee as perfect and dependable energy sources.

Thermal power is many times considered an elective energy source because of its low ozone harming substance outflows and potential to give reliable baseload power. Thermal energy stations utilize the course of atomic splitting to let energy out of nuclear cores. Be that as it may, the utilization of thermal power is joined by worries about security, radioactive garbage removal, and the potential for atomic weapons multiplication. Thus, thermal power stays a subject of discussion and debate.

The quest for elective energy sources is driven not just by the need to battle environmental change yet additionally by the longing to upgrade energy security. Conventional energy sources, like oil and flammable gas, are frequently imported from politically temperamental locales, prompting worries about supply interruptions and international strains. Elective energy sources, especially those that are locally created, offer a method for decreasing reliance on petroleum derivatives and the related dangers of supply interruptions.

Monetary turn of events and occupation creation are extra drivers in the journey for elective energy sources. As the sustainable power area develops, it produces business open doors and animates financial development. Sun powered and wind enterprises, specifically, have encountered huge work development lately. The turn of events and execution of new advancements in the elective energy area make occupations in assembling, establishment, upkeep, and innovative work.

Besides, the journey for elective energy sources is intently attached to energy effectiveness. Further developing energy effectiveness in businesses, structures, and transportation decreases by and large energy interest and the requirement for new energy sources. It additionally offers financial advantages, as energy-proficient advancements can lessen working expenses and increment seriousness. Energy proficiency is a basic piece of the more extensive technique to progress to cleaner energy sources.

The change to elective energy sources isn't without its difficulties. While these sources hold extraordinary commitment, they additionally face critical obstacles regarding innovation, framework, strategy, and public insight.

One of the key specialized difficulties is energy stockpiling. Numerous elective energy sources, for example, sun powered and wind, are irregular, creating power just when the sun is sparkling or the breeze is blowing. Successful energy stockpiling arrangements are vital for span the holes between energy age and utilization. Batteries, similar to those utilized in electric vehicles, are a promising stockpiling choice, however they need to turn out to be more productive, savvy, and harmless to the ecosystem.

Network reconciliation is another specialized test. The current electrical network was intended to oblige concentrated power age from petroleum derivatives. As elective energy sources, especially disseminated ones like roof sunlight powered chargers, become more predominant, the network should adjust to deal with these variable wellsprings of energy. Matrix modernization and the improvement of brilliant lattices are vital for productive and dependable energy conveyance.

Strategy and guideline assume a crucial part in the reception of elective energy sources. Government motivating forces, endowments, and commands can drive interest in clean energy advancements and establish a positive market climate. Alternately, conflicting or negative strategies can smother progress. The political and administrative scene changes from one country to another, which can influence the speed of progress to elective energy sources.

Public discernment and acknowledgment are additionally critical variables. While there is developing consciousness of the significance of clean energy, difficulties, for example, the stylish effect of wind turbines, worries about property estimations close to sun powered ranches, and resistance to new energy foundation can make impediments to the extension of elective energy sources. Successful correspondence and local area commitment are fundamental for address these worries.

One more test in the progress to elective energy sources is the requirement for a different energy blend. No single elective energy source can fulfill all energy needs for all areas and purposes. A blend of sources is important to guarantee unwavering quality and versatility. Adjusting this blend while considering nearby circumstances and limitations is a complicated errand.

7.2 The rise of renewable energy and electric vehicles.

In the 21st 100 years, the world is seeing a critical shift towards cleaner and more practical wellsprings of energy and transportation. The ascent of sustainable power and electric vehicles is a reaction to the developing worries over environmental change, energy security, and natural maintainability. This extraordinary shift is reshaping the way that we produce power and move individuals and products, with the possibility to lessen fossil fuel byproducts, improve energy autonomy, and alter the transportation area.

The mission for environmentally friendly power has picked up speed as the effects of environmental change have become progressively obvious. Customary wellsprings of energy, like coal, oil, and petroleum gas, have been the essential drivers of ozone depleting substance emanations, which add to a dangerous atmospheric devation and its related results, including increasing temperatures, outrageous climate occasions, and ocean level ascent. Perceiving the requirement for choices, environmentally friendly power sources have arisen as an economical and harmless to the ecosystem arrangement.

Sunlight based power is at the very front of the sustainable power transformation. Sun based energy tackles the force of the sun, which gives a plentiful and basically limitless wellspring of energy. Sun powered chargers, otherwise called photovoltaic cells, convert daylight into power. The innovation has progressed quickly as of late, making sun oriented power progressively open and savvy. Sun oriented establishments can be tracked down on private housetops, in monstrous sun based ranches, and in metropolitan settings, giving clean energy to control homes and organizations.

Wind energy is another conspicuous environmentally friendly power source that has seen astounding development. Wind turbines catch the motor energy of the breeze and convert it into power. Huge scope wind ranches, both coastal and seaward, have been created in locales with reliable breeze designs. Wind power enjoys the benefit of being harmless to the ecosystem and sustainable, yet it is dependent upon varieties in wind speed and bearing. Propels in turbine innovation and network joining have assisted make with winding energy a more solid wellspring of force.

Hydropower, or hydroelectric power, uses the energy of streaming water to create power. Dams and supplies are developed to control the progression of water, which is guided through turbines to create electrical power. Hydropower is a deep rooted wellspring of environmentally friendly power and has been a critical supporter of worldwide power age for a long time. It is dependable, delivers no immediate ozone depleting substance discharges, and can be acclimated to satisfy fluctuating energy needs. Notwithstanding, its natural effects, like the disturbance of oceanic biological systems and uprooting of networks, have provoked a quest for additional maintainable ways to deal with hydropower.

Geothermal energy takes advantage of the World's interior intensity by utilizing heat siphons to move warmth starting from the earliest stage structures or by outfitting the steam and boiling water found underneath the World's surface to produce power. Geothermal power plants are regularly situated in districts with critical geothermal action, like fountains and underground aquifers. This sustainable power source is profoundly dependable, accessible every minute of every day, and delivers extremely low ozone depleting substance outflows. As innovation propels, the potential for tackling geothermal energy grows to areas past those with clear geothermal elements.

Biomass energy gets from natural materials, like wood, farming buildups, and even green growth. Biomass can be scorched to create heat or changed over into biofuels, as biodiesel or ethanol, to drive vehicles. While biomass is considered inexhaustible in light of the fact that the plants used to deliver it tends to be regrown, it can in any case add to ecological issues, like deforestation and rivalry for land and assets. Cutting edge innovations are arising to alleviate these difficulties and make biomass energy more supportable.

Flowing and wave energy are generally new and promising types of environmentally friendly power. Tides and waves, driven by the gravitational powers of the moon and the sun, offer a reliable and unsurprising wellspring of energy. Flowing energy is bridled by catching the development of water during elevated and low tides, while wave energy is caught by gadgets that move with the movement of the waves. These advances are still in the beginning phases of improvement yet hold guarantee as perfect and dependable energy sources.

The ascent of sustainable power is a reaction to the squeezing need to address environmental change and diminish our carbon impression. Sustainable power sources offer a few key benefits:

Decreased Fossil fuel byproducts: Not at all like petroleum derivatives, sustainable power sources produce practically no ozone harming substance discharges. Sun based, wind, and hydroelectric power, specifically, have a negligible carbon impression, making them critical in the battle against environmental change.

Energy Autonomy: Environmentally friendly power sources are much of the time homegrown, which diminishes dependence on imported petroleum derivatives and improves energy security. This lessens weakness to worldwide energy market vacillations and international struggles.

Feasible Asset: Environmentally friendly power sources are, by definition, economical and limitless. Sun based and wind power, for instance, depend on the constant and plentiful accessibility of daylight and wind, making them a reliable wellspring of energy.

Work Creation: The development of the environmentally friendly power area has set out business open doors in assembling, establishment, upkeep, and innovative work. This occupation creation invigorates financial development and cultivates an environmentally friendly power labor force.

Mechanical Progressions: As the environmentally friendly power area extends, it drives innovative advancements and cost decreases. Progresses in sunlight powered charger productivity, wind turbine plan, and energy stockpiling are making environmentally friendly power more available and savvy.

Framework Flexibility: Appropriated energy sources, like roof sunlight based chargers, can improve network versatility and lessen the gamble of broad blackouts. Confined power age can give energy during framework disappointments or catastrophic events.

Natural Advantages: Sustainable power sources have a lower ecological effect contrasted with petroleum derivatives. They lessen air and water contamination, save biological systems, and limit natural surroundings disturbance.

Electric vehicles (EVs) have turned into a fundamental part of the progress to cleaner transportation. The ascent of EVs is driven by the need to lessen ozone harming substance emanations, improve air quality, and diminishing our reliance on oil. As conventional gas powered motor vehicles (ICEVs) are a huge wellspring of fossil fuel byproducts, especially in the transportation area, the reception of electric vehicles addresses a change in perspective with significant ecological and monetary ramifications.

The essential benefit of electric vehicles is their capability to lessen ozone depleting substance emanations. While the natural effect of an EV relies upon the wellspring of power utilized for charging, in locales with a critical portion of sustainable power in the matrix, EVs can deliver significantly lower fossil fuel byproducts contrasted with their gas or diesel partners. This decrease in emanations is basic for alleviating environmental change and further developing air quality in metropolitan regions.

Battery electric vehicles (BEVs) and module cross breed electric vehicles (PHEVs) are the two fundamental kinds of electric vehicles. BEVs are completely electric, depending exclusively on electric power for impetus. They have bigger battery limits and longer all-electric reaches, making them reasonable for day to day drives and longer excursions. PHEVs join an electric engine with a gas powered motor, permitting drivers to switch among electric and gas power. This adaptability tends to worries about range impediments and the accessibility of charging foundation.

The ecological advantages of electric vehicles stretch out past decreased ozone harming substance emanations. EVs additionally produce less air contamination, as they don't emanate tailpipe toxins like nitrogen oxides (NOx) and particulate matter (PM) related with ICEVs. Cleaner air quality can prompt better general wellbeing results, with less instances of respiratory sicknesses and related wellbeing costs.

Moreover, electric vehicles add to energy freedom by diminishing dependence on oil imports. Oil reliance is a critical worry in numerous nations, as it can prompt weaknesses connected with worldwide oil market changes and international pressures. Electric vehicles depend on power, which can be created from a different blend of fuel sources, including renewables, subsequently upgrading energy security.

7.3 Challenges and opportunities in transitioning away from oil.

The world's reliance on oil has for some time been a principal quality of current culture. Oil drives our transportation frameworks, gives the unrefined substances to endless items, and assumes a pivotal part in worldwide energy creation. Be that as it may, this dependence on oil accompanies a large group of difficulties, going from natural worries to international contentions.

As we face the need to resolve these issues, progressing away from oil turns into an objective. This progress presents a complicated scene of difficulties and open doors that will shape the eventual fate of energy, the climate, and worldwide economies.

Quite possibly of the most squeezing challenge in changing away from oil is alleviating the ecological effect of petroleum product utilization. The consuming of oil for transportation and energy creation discharges carbon dioxide (CO_2) and different poisons into the air, adding to environmental change and air contamination. To resolve this issue, the world is progressively going to option and cleaner energy sources, like flammable gas, environmentally friendly power, and atomic power. Nonetheless, these changes are not without their difficulties.

The shift towards petroleum gas is viewed as a cleaner choice to oil. It produces less CO_2 emanations and air poisons when consumed. Notwithstanding, the extraction and transportation of gaseous petrol can bring about methane spills, which are intense ozone depleting substances. Moreover, flammable gas is a limited asset, and its drawn out ecological advantages are reliant upon successful methane catch and the improvement of economical extraction rehearses.

Environmentally friendly power sources, for example, sun oriented and wind power, offer a promising way away from oil-based energy. These sources are sustainable, bountiful, and produce no immediate ozone harming substance outflows during power age. However, their irregular nature and reliance on energy capacity advances present difficulties in giving predictable power supply. Besides, progressing to renewables requires significant framework ventures, for example, building wind ranches and sunlight based establishments, and further developing lattice coordination to oblige disseminated energy sources.

Atomic power is one more choice to oil that gives a low-carbon energy source. Atomic plants produce power without direct CO2 outflows and can give baseload power. Be that as it may, the atomic business faces difficulties connected with security, radioactive garbage removal, and the potential for atomic weapons multiplication. Public insight and administrative obstacles can likewise thwart the extension of thermal power.

Progressing away from oil in the transportation area is another basic test. Oil fills most of vehicles internationally, and this dependence adds to high outflows of CO2, as well as poisons that mischief air quality. Electric vehicles (EVs) have built up forward momentum as a cleaner elective, yet the development of EV reception depends on extending charging framework, further developing battery innovation, and tending to worries about range impediments. Furthermore, the transportation area incorporates substantial and long stretch vehicles for which charge is really difficult.

One more huge test in changing away from oil is the effect on economies that are vigorously subject to oil creation. Numerous nations depend on oil as a significant wellspring of income and business.

The shift away from oil can prompt financial disturbances, as declining interest for oil can prompt lower costs, diminished government incomes, and employment misfortunes in the oil business. State run administrations in oil-creating countries need to broaden their economies, put resources into elective ventures, and foster techniques for changing their labor forces to new open doors.

International worries are additionally interwoven with the difficulties of changing away from oil. The worldwide energy scene has for quite some time been molded by the essential interests of oil-creating countries, prompting clashes and strains over admittance to and control of oil holds. As nations diminish their reliance on oil, these international elements might move, setting out new open doors for tact and decreasing the potential for energy-related clashes. Be that as it may, new difficulties might emerge as countries vie for strength in the environmentally friendly power and innovation areas.

The requirement for energy security is a central worry that impacts the progress away from oil. Dependence on oil imports can make nations powerless against supply disturbances and market variances. Changing to more different and locally delivered energy sources, like environmentally friendly power, can improve energy security by decreasing reliance on unfamiliar oil and making energy supply stronger to shocks. Energy capacity advances, further developed lattice framework, and productive energy the board likewise assume essential parts in guaranteeing energy security.

One more test in progressing away from oil is the need to create and execute powerful approaches and guidelines. Government motivations, endowments, and commands can drive interest in clean energy innovations and establish a good market climate. Notwithstanding, conflicting or troublesome arrangements can smother progress. The political and administrative scene differs from one country to another, influencing the speed of the change. Facilitated endeavors are required at the public and global levels to lay out clear and steady strategy systems.

Resolving the issue of abandoned resources is likewise a test in the change away from oil. Abandoned resources allude to interests in oil-related framework, like pipelines, processing plants, and boring gear, which might lose esteem as the interest for oil diminishes. Tracking down elective purposes or reusing these resources presents financial and ecological difficulties. Anticipating the fate of these resources is fundamental to limit financial disturbance and ecological mischief.

In spite of the horde challenges related with progressing away from oil, there are various open doors for financial development, ecological improvement, and worldwide steadiness. These open doors are firmly connected to the turn of events and reception of clean energy advances and the reshaping of energy and transportation frameworks.

The progress to option and cleaner energy sources offers significant financial open doors. The environmentally friendly power area, including sun based, wind, and hydropower, has encountered huge development and occupation creation. Interests in environmentally friendly power tasks, innovative work, and assembling are driving financial turn of events and cultivating a gifted labor force. Additionally, the electric vehicle industry has seen striking development, making position in vehicle fabricating, battery creation, and charging foundation improvement.

Mechanical headways are a main impetus behind the progress away from oil. Sustainable power advancements, like sunlight based chargers and wind turbines, have become more proficient and financially savvy. Battery innovation has improved, taking into account better energy stockpiling arrangements. High level lattice framework and shrewd network advances are upgrading energy dissemination and the executives. The improvement of energy-productive machines and transportation choices, similar to high velocity rail and electric transports, adds to diminishing energy utilization.

Ecological advantages are a central open door in the progress away from oil. Cleaner energy sources and electric vehicles produce less ozone depleting substance outflows, which are fundamental to relieving environmental change. Further developed air quality, coming about because of diminished contamination from the transportation area, prompts better general wellbeing and personal satisfaction. Protecting biological systems and diminishing natural surroundings disturbance, which frequently go with non-renewable energy source extraction, are fundamental ecological benefits.

Energy security is a basic open door that accompanies the progress away from oil. Broadening energy sources and lessening reliance on unfamiliar oil imports make nations stronger to supply interruptions and market changes. The advancement of homegrown energy assets, for example, renewables and flammable gas, improves energy freedom and diminishes the dangers related with international pressures and clashes over oil holds.

The change away from oil presents conciliatory and international open doors. Diminishing the essential significance of oil can reduce the motivation for energy-related clashes. As nations coordinate to foster environmentally friendly power foundation and innovation, worldwide relations might be reshaped around clean energy exchange and joint effort. Strategy can assume a fundamental part in molding the worldwide energy scene and cultivating global collaboration.

Public discernment and cultural qualities offer open doors in the change away from oil. Expanding attention to the natural effect of petroleum products and the advantages of clean energy arrangements can drive buyer inclinations and conduct. Cultural qualities that focus on maintainability, natural obligation, and moral contemplations can speed up the reception of clean energy advancements and the shift away from oil.

The change to elective energy sources and transportation choices additionally gives potential open doors to advancement and business venture. New businesses and business visionaries are driving the advancement of new innovations, plans of action, and answers for address the difficulties of the change.

Chapter 8

The Modern Oil Industry

The cutting edge oil industry is an immense and complex worldwide endeavor that assumes a significant part in fueling economies, forming international affairs, and affecting the climate. It has made considerable progress from its unassuming starting points in the late nineteenth century when oil was first found as an important energy asset. Today, the business incorporates investigation, penetrating, extraction, refining, transportation, dissemination, and innumerable subordinate exercises that give energy and unrefined substances to many applications. Nonetheless, this monstrous and persuasive industry isn't without its difficulties and debates.

The Starting points and Development of the Oil Business

The oil business, frequently alluded to as the petrol business, has its foundations in the US, with the revelation of business oil stores in Pennsylvania during the nineteenth 100 years. This revelation prompted the foundation of the main business oil well, the Drake Well, in 1859. The outcome of this very much denoted the start of a fast extension of the oil business, as new fields were found, and new innovations were created to extricate and refine oil based commodities.

One of the earliest and most huge players in the oil business was John D. Rockefeller, who established Standard Oil in 1870. Standard Oil immediately turned into a predominant power in the business, controlling the creation, refining, and dissemination of oil. In any case, this predominance prompted enemy of trust activities, eventually bringing about the separation of Standard Oil in 1911 into more modest, free organizations.

The twentieth century saw the proceeded with development of the worldwide oil industry, with the Center East arising as a significant wellspring of oil creation. The revelation of huge oil holds in nations like Saudi Arabia and Iran prompted the foundation of worldwide oil organizations and the advancement of the cutting edge oil exchange. The Association of the Oil Sending out Nations (OPEC) was shaped in 1960 to facilitate oil creation and valuing among its part nations, giving them more command over their oil assets.

The Oil Business Today

Today, the cutting edge oil industry is a complicated organization of worldwide partnerships, public oil organizations, and a wide cluster of specialist co-ops. It traverses the globe, with significant oil-creating locales including the Center East, North America, Latin America, Africa, and Asia. These locales are home to immense oil saves, and organizations and state run administrations work to concentrate and product this important asset.

The investigation and penetrating for oil include state of the art innovation, from seismic imaging to remote ocean boring and pressure driven cracking (deep oil drilling). These strategies are utilized to find and concentrate oil from various sources, including ordinary oil fields, seaward holds, and unusual sources like shale endlessly oil sands.

Whenever oil is extricated, it should be refined to create a scope of items, including gas, diesel, fly fuel, and petrochemicals. Processing plants are complicated offices that interaction unrefined petroleum into these different items, while likewise attempting to fulfill ecological guidelines and diminish the natural effect of their tasks.

The transportation and dispersion of oil are likewise basic parts of the business. Oil should be moved from creation locales to treatment facilities and from processing plants to appropriation focuses and buyers. Pipelines, big haulers, and rail transport are completely used to move oil starting with one area then onto the next.

The Worldwide Effect of the Oil Business

The cutting edge oil industry significantly affects worldwide economies and international relations. As an essential wellspring of energy, oil is a major driver of financial development and improvement. It energizes transportation frameworks, powers ventures, and fills in as a feedstock for the creation of endless customer products. Oil costs can impact expansion rates, exchange adjusts, and the general strength of economies around the world.

The significance of oil is especially obvious in the transportation area, where gas and diesel fuel are the essential wellsprings of energy for vehicles, trucks, and planes. The accessibility and reasonableness of oil straightforwardly influence the expense of transportation, which, thus, influences the expense of labor and products. Oil value variances can set out financial difficulties and open doors, influencing businesses and shoppers the same.

Geopolitically, the oil business plays had a critical impact in molding global relations and clashes. Oil-creating nations frequently use impressive impact because of their command over fundamental energy assets. The worldwide interest for oil has prompted discretionary endeavors and, on occasion, military mediations to tie down admittance to oil saves and delivery courses. This has made both vital collusions and pressures between countries.

The cutting edge oil industry additionally faces concerns connected with natural maintainability and environmental change. The ignition of non-renewable energy sources, including oil, discharges ozone harming substances like carbon dioxide into the climate, adding to a dangerous atmospheric devation and its related effects. The ecological outcomes of oil extraction and transportation, including oil slicks and territory disturbance, are likewise subjects of concern.

The Difficulties and Discussions of the Oil Business

The advanced oil industry isn't without its difficulties and discussions. These issues envelop natural, social, monetary, and political aspects, and they frequently flash public discussion and administrative reactions.

Environmental Change and Ecological Effect: Perhaps of the most squeezing challenge confronting the oil business is its commitment to environmental change. The consuming of oil for energy is a significant wellspring of fossil fuel byproducts, which are an essential driver of an Earth-wide temperature boost. The business faces calls to lessen its carbon impression and put resources into cleaner energy sources.

Oil slicks and Ecological Fiascos: Mishaps, for example, oil slicks, represent a huge natural danger. The Deepwater Skyline oil slick in the Bay of Mexico in 2010 and the Exxon Valdez spill in 1989 are instances of disastrous occasions with annihilating ecological results.

Environment Disturbance and Biodiversity Misfortune: The extraction and transportation of oil can upset biological systems, hurt untamed life, and lead to living space misfortune. Growing oil foundation frequently infringes on normal regions, influencing both earthly and marine conditions.

Asset Exhaustion: The consumption of limited oil saves raises worries about future energy security. As effectively available stores are depleted, the business should put resources into new advances and investigate flighty sources, for example, oil sands and shale oil.

Financial Weakness: Oil-subordinate economies are defenseless against changes in oil costs. An unexpected drop in costs can prompt financial unsteadiness and government income setbacks, as found in some oil-delivering countries during times of low oil costs.

Political Struggles and International Pressures: The mission for command over oil holds has energized political contentions and international strains. Contest for admittance to oil assets can prompt debates and, surprisingly, military intercessions.

Social and Monetary Imbalance: The advantages of oil abundance are not generally conveyed equitably. In some oil-creating nations, abundance abberations endure, and nearby networks may not receive the financial rewards of oil extraction, prompting social agitation.

Common liberties and Work Concerns: Work conditions in the oil business, especially in a few emerging nations, have raised worries about laborer security, fair wages, and denials of basic freedoms.

Administrative and Moral Difficulties: The oil business is dependent upon complex administrative conditions, including natural guidelines and security principles. Moral contemplations, like corporate social obligation and moral speculation, are progressively significant for oil organizations.

Open doors and Reactions

Tending to the difficulties and discussions of the cutting edge oil industry requires a diverse methodology that includes legislatures, the confidential area, common society, and worldwide associations. A portion of the open doors and reactions to these difficulties include:

Progress to Clean Energy: The change to cleaner and sustainable power sources is a critical reaction to the test of environmental change. Interest in sunlight based, wind, and hydroelectric power, as well as examination into energy capacity advances, is diminishing the carbon impression of the energy area.

Natural Guidelines: Severe ecological guidelines and wellbeing principles assist with moderating the business' effect on the climate. Legislatures and worldwide associations assume a basic part in setting and implementing these guidelines.

Preservation and Biodiversity Drives: Endeavors to secure and reestablish normal living spaces and biodiversity can assist with moderating the ecological effect of oil extraction and transportation.

Supportable Asset The board: Embracing manageable practices in oil extraction and diminishing waste can assist with drawing out the existence of existing stores and limit the natural effect.

Monetary Expansion: Oil-subordinate economies can enhance by putting resources into different areas, like innovation, assembling, and the travel industry. This decreases weakness to oil cost changes.

Social Obligation: Oil organizations are progressively expected to participate in corporate social obligation drives, helping neighborhood networks, and tending to work and basic liberties concerns.

Environmentally friendly power Speculations: Many oil organizations are broadening their portfolios by putting resources into environmentally friendly power projects, perceiving the developing significance of clean energy sources.

Mechanical Advancement: The improvement of new innovations, for example, carbon catch and use, can assist with decreasing the fossil fuel byproducts related with oil creation and use.

8.1 The current state of the global oil industry.

The worldwide oil industry is at a junction, confronting a mix of difficulties and potential open doors that are reshaping its scene. As quite possibly of the most basic area in the worldwide economy, the condition of the oil business has extensive ramifications for energy security, monetary strength, ecological manageability, and international elements. In this time of speeding up change, understanding the present status of the worldwide oil industry is fundamental for settling on informed conclusions about its future and the worldwide energy progress.

Oil Creation and Stores

Oil stays an imperative wellspring of energy for the world, with a great many barrels removed and consumed day to day. As of the latest information accessible, the worldwide oil creation in 2021 surpassed 90 million barrels each day (bpd). While this creation level is significant, it is basic to perceive that the worldwide oil industry faces the test of recharging stores to guarantee a drawn out supply. Demonstrated worldwide oil holds are assessed to be around 1.7 trillion barrels, which is comparable to roughly 50 years of creation at the 2021 rate.

The conveyance of oil saves is exceptionally lopsided, with few nations holding the greater part. The Center East, eminently Saudi Arabia, Iran, Iraq, and the Unified Bedouin Emirates, has the biggest portion of worldwide stores. Venezuela, Russia, and the US likewise have critical stores. Conversely, numerous nations, especially in Europe and Africa, have restricted homegrown holds and depend on imports to satisfy their oil need.

Oil Costs and Market Elements

Oil costs are a critical sign of the condition of the worldwide oil industry, mirroring the exchange of organic market, international elements, financial circumstances, and worldwide energy strategies. The Coronavirus pandemic significantly affected oil costs, with the interest shock brought about by lockdowns and travel limitations prompting a lofty drop in costs in mid 2020. The cost of West Texas Transitional (WTI) unrefined petroleum momentarily turned negative in April 2020, a memorable occasion mirroring an outrageous oversupply circumstance.

Since the underlying shock of the pandemic, oil costs have bounced back, despite the fact that they remain affected by a few elements. Creation cuts by the Association of the Oil Trading Nations (OPEC) and its partners, including Russia (by and large known as OPEC+), play had an impact in balancing out costs. The energy progress, which has provoked interests in environmentally friendly power sources and expanded energy proficiency, has likewise presented a component of vulnerability about future oil interest.

International pressures and occasions in oil-creating areas keep on affecting oil costs. Disturbances in oil supply, whether because of international contentions, cataclysmic events, or specialized disappointments, can prompt cost spikes. Moreover, worldwide energy strategies, for example, carbon valuing and emanations decrease targets, can possibly impact the drawn out direction of oil interest and costs.

Energy Progress and Clean Energy Drives

One of the characterizing highlights of the present status of the worldwide oil industry is the speeding up shift toward cleaner and more economical energy sources. The earnest need to battle environmental change and lessen ozone harming substance discharges has prompted a worldwide push for decarbonization, influencing the job and viewpoint for oil.

Nations and locales are defining aggressive objectives to change away from petroleum products, including oil, and are effectively putting resources into sustainable power sources, for example, sun based, wind, and hydroelectric power. State run administrations are carrying out approaches to advance electric vehicles (EVs) and improve energy proficiency in transportation, structures, and businesses. These drives are pointed toward diminishing oil utilization and its related natural effects.

The oil business itself isn't invulnerable to the change. Many oil organizations, perceiving the changing energy scene, are differentiating their portfolios by putting resources into sustainable power ventures and advancements. They are additionally investigating amazing open doors in carbon catch and usage (CCU) and carbon sequestration, expecting to decrease the fossil fuel byproducts related with oil creation and refine their natural impression.

Worldwide energy strategies and guidelines are progressively focusing on fossil fuel byproducts and making way for a low-carbon energy future. Carbon estimating instruments, emanations decrease targets, and sustainable power orders are becoming normal devices for legislatures to drive the progress. Financial backers and shoppers arc likewise showing a developing revenue in reasonable and mindful speculations, which are coming down on oil organizations to adjust to changing business sector inclinations.

Electric Vehicles (EVs) and Transportation Area

The transportation area, liable for a huge portion of worldwide oil interest, is going through a change. Electric vehicles (EVs) are picking up speed and address a promising option in contrast to gas powered motor vehicles (ICEVs) fueled by gas and diesel.

The electric vehicle market is extending quickly, with automakers across the globe presenting new EV models. Further developed battery innovation, improved charging foundation, and government motivations are making EVs more open and interesting to buyers. Numerous nations have set focuses to progressively get rid of ICEVs and advance the reception of EVs.

The development of EVs can possibly lessen oil utilization in the transportation area essentially. Notwithstanding, the speed of this change shifts from one locale to another, and the improvement of electric charging foundation stays a test. Hard core and long stretch vehicles, like trucks and ships, present remarkable difficulties in jolt, and elective powers, like hydrogen, are being investigated as possible arrangements.

Ecological and Environment Concerns

Ecological worries are fundamental to the present status of the worldwide oil industry. The consuming of oil for energy creation is a significant wellspring of carbon dioxide (CO2) outflows, which add to an Earth-wide temperature boost and environmental change. This has provoked expanded examination of the business' ecological effect and requires a decrease in fossil fuel byproducts.

Oil organizations are confronting developing strain to work on their ecological practices and diminish their carbon impression. Systems to address these worries incorporate carbon catch and use (CCU), where CO2 discharges are caught and utilized for different purposes, like upgraded oil recuperation or the creation of manufactured energizes. Carbon sequestration, which includes putting away CO2 in land developments, is likewise being investigated as a method for relieving discharges.

Natural guidelines and outflows decrease targets are influencing the oil business' tasks. Organizations are putting resources into innovation and practices to decrease methane outflows, limit erupting, and further develop energy proficiency in their tasks. The business is likewise investigating the capability of biofuels, which are gotten from sustainable assets and can possibly diminish ozone harming substance emanations.

International Elements

The international scene keeps on assuming a huge part in the worldwide oil industry. Oil-creating nations and areas hold key significance because of their command over essential energy assets. International strains, clashes, and unions shape admittance to oil stores and energy supply courses.

The Center East remaining parts a point of convergence of international elements in the oil business. The locale is home to huge oil holds and has been a key member in OPEC's endeavors to impact oil creation and estimating. International pressures in the Center East, including those connected with Iran, Saudi Arabia, and Iraq, have generally affected oil costs and supply dependability.

The US has arisen as an unmistakable oil maker, with the development of shale oil and gas creation prompting expanded energy freedom. Be that as it may, the US's oil industry is likewise impacted by worldwide elements, as it imports and commodities oil, influencing worldwide oil markets. The country's energy arrangements, including guidelines and creation levels, can shape the energy scene both locally and internationally.

Notwithstanding customary international pressures, the worldwide energy change is presenting new elements. As nations lessen their dependence on oil and shift toward cleaner energy sources, the energy scene is developing, prompting new open doors and difficulties in global relations.

Financial Ramifications

The condition of the worldwide oil industry has significant financial ramifications. The cost of oil straightforwardly influences economies around the world, as it impacts the expense of transportation, creation, and buyer merchandise. The accessibility of reasonable oil is fundamental for monetary steadiness and development.

Oil-subordinate economies, frequently found in locales with critical oil holds, are defenseless against variances in oil costs. Unexpected drops in costs can prompt financial shakiness, decreased government incomes, and employment misfortunes in the oil business. Nations with restricted broadening in their economies are especially vulnerable to these cost variances.

On the other hand, high oil costs can prompt expanded incomes for oil-creating countries, permitting them to put resources into foundation, social projects, and monetary turn of events. These incomes can likewise make monetary space for state run administrations to explore financial difficulties and put resources into broadening their economies.

The job of oil in the worldwide economy is additionally confounded by the changing elements of the energy progress. Interests in environmentally friendly power and clean advances are reshaping the energy area, setting out new financial open doors and difficulties. This change can bring about the redistribution of capital, positions, and assets across different businesses, influencing monetary development and work.

8.2 Technological advancements in oil exploration and extraction.

The oil and gas industry has for some time been at the front line of mechanical development. From the beginning of penetrating the principal business oil wells to the present state of the art investigation and extraction techniques, innovation has persistently pushed the limits of what's conceivable in the quest for and creation of hydrocarbons. These progressions have expanded the business' effectiveness as well as tended to complex difficulties, including getting to beforehand undiscovered stores, lessening ecological effect, and further developing security principles.

Oil Investigation Methods

Oil investigation is a basic initial phase during the time spent finding and getting to hydrocarbon saves. Throughout the long term, the business has fostered a scope of imaginative methods and innovations to pinpoint the area of oil and gas stores underneath the World's surface.

Seismic Imaging: Seismic overviews are a crucial device in oil investigation. This method includes creating and recording sound waves that movement through the World's subsurface and return quickly to the surface. By breaking down the manner in which these waves bounce off underground stone layers, geophysicists can make definite 3D pictures of subsurface designs and distinguish potential hydrocarbon repositories. Progressions in seismic innovation have prompted higher-goal pictures and more precise expectations of supply properties.

Gravity and Attractive Reviews: Gravity and attractive overviews are utilized to identify subsurface varieties in thickness and attractive properties. These studies can assist with recognizing the presence of topographical designs that might be related with oil and gas supplies. Present day instrumentation and information investigation strategies have improved the precision and proficiency of these reviews.

Electromagnetic Strategies: Electromagnetic techniques include estimating the electrical conductivity of subsurface rocks. These techniques are particularly valuable for recognizing hydrocarbon repositories in flighty developments, like shale. High level electromagnetic studies can give definite data about the creation and liquid substance of subsurface arrangements.

Remote Detecting and Satellite Innovation: Satellite symbolism and remote detecting advancements have upset how oil investigation is led. These advances empower the observing of surface elements and land changes over enormous regions. They are especially important for distinguishing potential investigation focuses in remote or unavailable locales.

AI and Information Examination: AI and information examination assume a significant part in handling and deciphering huge measures of geographical and geophysical information. These advances can reveal examples, connections, and irregularities in information, helping geoscientists in settling on additional educated conclusions about investigation targets.

Penetrating and Extraction Innovations

When an oil hold is recognized, the following test is to separate the hydrocarbons proficiently and securely. Mechanical headways in penetrating and extraction have altogether worked on the business' capacities in getting to and recuperating oil and gas assets.

Directional Boring: Directional penetrating permits drillers to adjust the course and direction of a wellbore. This innovation is fundamental for getting to holds that are not straightforwardly beneath the boring site. Directional boring strategies, for example, flat penetrating, empower the extraction of oil and gas from a more extensive region of the supply, expanding creation rates.

Water driven Cracking (Deep oil drilling): Pressure driven breaking, normally alluded to as deep earth drilling, is a strategy used to invigorate the progression of hydrocarbons from whimsical repositories, like shale and tight sandstone. It includes infusing water, sand, and synthetic substances at high strain to make breaks in the stone and delivery caught oil and gas. Propels in deep earth drilling innovation have made it conceivable to remove hydrocarbons from beforehand uneconomical developments.

Overseen Strain Penetrating (MPD): Oversaw Tension Boring is an innovation that considers better control of wellbore strain during penetrating tasks. This improves security, lessens the gamble of victories, and increments boring proficiency, especially in testing developments.

Boring apparatus Innovation: Present day bores are furnished with cutting edge materials and plans that improve penetrating effectiveness and sturdiness. PDC (polycrystalline jewel reduced) bores, for instance, offer better cutting execution and toughness in testing developments.

Underbalanced Boring: Underbalanced penetrating includes boring with a liquid that has lower tension than the development pressure. This procedure limits development harm and works on well efficiency. Propels in underbalanced boring innovation have worked on well execution and supply recuperation.

Boring Computerization: Mechanization and mechanical technology are progressively utilized in penetrating tasks to upgrade accuracy and productivity. Robotized boring frameworks can constantly screen and change penetrating boundaries, decreasing the gamble of human blunder and further developing security.

Rig Innovations: Seaward penetrating apparatuses have seen critical mechanical progressions. Dynamic situating frameworks, high level victory preventers, and further developed security highlights have made seaward penetrating more secure and more effective. Ultra-deepwater and unforgiving climate penetrating innovations have stretched out the business' compass to already undiscovered assets.

Improved Oil Recuperation (EOR): Upgraded Oil Recuperation strategies are utilized to remove additional oil from supplies after essential and auxiliary recuperation techniques have been depleted. Advancements like carbon dioxide infusion, polymer flooding, and steam infusion have been created to increment recuperation rates and broaden the existence of mature fields.

Ecological Contemplations

The oil and gas industry is under expanding strain to limit its ecological effect. Mechanical progressions have been instrumental in tending to ecological worries related with oil investigation and extraction.

Decreased Outflows: Advances in penetrating and creation innovations have prompted diminished methane emanations during boring and water driven cracking. These enhancements are fundamental in alleviating the business' commitment to ozone depleting substance discharges.

Zero-Erupting Innovations: Gas erupting, a training that consumes off overabundance petroleum gas, is a huge wellspring of natural concern. Zero-erupting advances mean to catch and use this gas, decreasing waste and outflows.

Carbon Catch and Capacity (CCS): CCS innovations catch carbon dioxide discharges from modern cycles, including oil and gas creation, and store them underground. CCS is viewed as a crucial device in decreasing the business' carbon impression.

Water Reusing and Reuse: Water is fundamental in pressure driven breaking and penetrating tasks. High level water the executives advances empower the reusing and reuse of delivered and flowback water, diminishing freshwater utilization and wastewater removal.

High level Materials and Boring Liquids: Harmless to the ecosystem penetrating liquids and materials are created to lessen the effect of penetrating procedure on nearby biological systems and water assets.

Jolt of Seaward Activities: Seaward penetrating tasks are progressing to electric power frameworks to diminish the dependence on diesel generators, which discharge ozone harming substances. Jolt decreases discharges and works on functional effectiveness.

8.3 The future outlook for the petroleum industry.

The petrol business is at a basic point, confronting a quickly changing scene driven by monetary, ecological, and international powers. The business, which plays had a focal impact in fueling worldwide economies for north of a long time, is currently exploring a future set apart by vulnerabilities, difficulties, and open doors. In this conversation, we investigate the advancing standpoint for the petrol business and the variables that are molding its future.

Developing Energy Blend:

Quite possibly of the main variable influencing the eventual fate of the oil business is the continuous change in the worldwide energy blend. As worries about environmental change and ecological manageability escalate, there is a developing accentuation on progressing away from non-renewable energy sources, including oil. Legislatures, organizations, and shoppers are progressively focusing on elective energy sources, for example, renewables and petroleum gas, to decrease fossil fuel byproducts and address long haul energy needs.

Environmentally friendly power Progress:

Environmentally friendly power sources, including sunlight based, wind, and hydropower, have seen momentous development as of late. Progresses in innovation, falling expenses, and government motivators have made environmentally friendly power a cutthroat and supportable option in contrast to customary petroleum products. This change towards renewables is driven by a few key variables:

Natural Worries: The need to relieve the effects of environmental change and lessen ozone depleting substance discharges has placed tension on states and ventures to embrace cleaner energy sources. Sustainable power offers a lower carbon impression contrasted with non-renewable energy sources, pursuing it a favored decision in a carbon-obliged world.

Energy Security: Differentiating energy sources with renewables improves energy security by lessening dependence on oil imports. Nations with plentiful sustainable assets can saddle their homegrown energy potential, diminishing their weakness to worldwide energy market vacillations.

Cost Intensity: The declining expenses of sustainable power innovations, like sunlight based chargers and wind turbines, have made them progressively reasonable. This cost intensity has driven the reception of renewables as a standard energy source.

Mechanical Headways: Progressing progressions in energy capacity and network joining advances are tending to the irregularity and unwavering quality issues related with renewables, making them more trustworthy for satisfying energy needs.

Strategy and Guidelines: Legislatures all over the planet are executing arrangements to advance sustainable power reception. These incorporate sustainable portfolio norms, feed-in duties, and carbon estimating systems, all of which boost the turn of events and use of clean energy.

The rising conspicuousness of sustainable power sources represents a test to the oil business. A shift away from oil as the predominant energy source could prompt a decreased interest for oil based goods, especially in the transportation area, where gas and diesel energizes have generally been the essential options. Thus, oil organizations are enhancing their portfolios to incorporate environmentally friendly power projects, like breeze and sunlight based ranches, to stay cutthroat and adjust to changing business sector elements.

Electric Vehicles (EVs):

The transportation area is a huge purchaser of oil based commodities, representing a significant piece of worldwide oil interest. Electric vehicles (EVs), fueled by power from the framework, are turning into an appealing option in contrast to gas powered motor vehicles (ICEVs) burning gas or diesel. A few variables are driving the development of electric portability:

Natural Advantages: EVs produce zero tailpipe discharges, which lines up with the push for cleaner and more reasonable transportation. Lessening air contamination and checking ozone harming substance discharges are convincing motivations to progress to electric vehicles.

Mechanical Progressions: Continuous headways in battery innovation are expanding the reach and reasonableness of EVs. As battery costs keep on declining, EVs are turning out to be more available to shoppers.

Energy Productivity: Electric vehicles are intrinsically more energy-effective than conventional ICEVs. This proficiency means lower working expenses and decreased energy utilization.

Government Motivators: Numerous states are offering impetuses to advance EV reception, for example, tax reductions, discounts, and admittance to high-inhabitance vehicle paths.

Customer Interest: Developing familiarity with ecological issues and the longing for feasible transportation choices are driving shopper interest in electric vehicles.

The ascent of EVs presents a test to the petrol business as it straightforwardly influences the interest for gas and diesel fills. Oil organizations are answering by investigating open doors in the electric vehicle charging foundation and battery innovation areas. Moreover, they are expanding into different types of versatility, for example, hydrogen energy units, which offer an option in contrast to conventional gas controlled vehicles.

Hydrogen as an Energy Transporter:
Hydrogen has acquired consideration as a flexible and clean energy transporter with the possibility to reshape the energy scene. Hydrogen can be delivered from different sources, including gaseous petrol, water electrolysis, and biomass, and it tends to be utilized as a fuel or feedstock in different areas. Its application in the petrol business incorporates the development of hydrogen fuel for vehicles and the utilization of hydrogen in modern cycles, like refining and petrochemical creation.

Green Hydrogen: Green hydrogen is created through water electrolysis, utilizing sustainable power sources to control the cycle. It is viewed as harmless to the ecosystem since it produces zero fossil fuel byproducts during creation. Green hydrogen can possibly supplant customary hydrogen creation techniques, which frequently depend on gaseous petrol.

Hydrogen Energy units: Hydrogen power devices are a promising innovation for different applications, including driving vehicles and giving reinforcement power in distant areas. Power device vehicles (FCVs) are an option to both ICEVs and battery-electric vehicles, as they offer longer ranges and more limited refueling times.

Modern Applications: Hydrogen is utilized in different modern cycles, like hydrocracking in processing plants, smelling salts creation, and steelmaking. The mix of hydrogen in these cycles can prompt diminished fossil fuel byproducts and worked on ecological execution.

Chapter 9

Oil's Enduring Legacy

Over the span of mankind's set of experiences, not many substances have applied such a significant and expansive impact as oil. It has, throughout the long term, become the backbone of our cutting edge industrialized world, forming economies, governmental issues, and social orders on a worldwide scale. Its effect is both inescapable and diverse, crossing from the mechanical wonders of the Modern Insurgency to the international struggles of the 21st hundred years. In this article, we will investigate the persevering through tradition of oil, looking at its authentic importance, its job in contemporary society, and the difficulties it models for what's in store.

The tale of oil is one of asset abuse and mechanical development. The underlying foundations of the cutting edge oil industry can be followed back to the mid-nineteenth century when the main business oil very much was bored in Pennsylvania, introducing another time of energy creation. Preceding this, oil was a to a great extent undiscovered asset, leaking normally from the World's surface. Be that as it may, the improvement of boring procedures and the revelation of huge oil holds made ready for an extreme change of the worldwide energy scene.

The Modern Unrest, a urgent period in mankind's set of experiences, was powered by the plentiful and reasonable energy source that oil gave. The approach of the steam motor, made conceivable by the usage of coal and later oil, upset transportation, assembling, and agribusiness. Trains, ships, and cars generally depended on this recently discovered energy source, moving society into a time of extraordinary development and thriving.

As the twentieth century unfolded, oil's significance just developed. The gas powered motor, which turned into the main thrust behind the ascent of the car business, was subject to oil for fuel. This change in transportation not just impacted the manner in which individuals lived and worked yet in addition had significant financial and social results. Suburbanization, for example, became conceivable as the auto permitted individuals to drive longer distances, reshaping metropolitan scenes and reclassifying the Pursuit of happiness.

Oil's importance stretches out past transportation; it supports the whole worldwide energy area. From power age to warming, oil plays had a critical impact in giving the energy that powers current life. Moreover, oil has been an indispensable feedstock for the petrochemical business, which creates a variety of items, from plastics to drugs. The sweeping effect of oil should be visible in the far and wide utilization of petrochemical-determined materials in our regular routines.

Geopolitically, oil has been a strong power, forming the elements of global relations. The mission for oil assets has been a driver of contentions and collusions, as countries tried to tie down their admittance to this significant item. The twentieth century saw oil's focal job in forming the international relations of the Center East, with the foundation of significant oil-delivering countries like Saudi Arabia and Iran. The control and conveyance of oil have frequently been at the core of worldwide questions and clashes, making it a basic part of worldwide epic showdowns.

Oil has additionally been instrumental in driving monetary development and advancement. Oil-rich nations enjoy experienced huge monetary benefits, as their regular assets turned into a wellspring of income and impact. Countries, for example, the Unified Bedouin Emirates and Norway have actually saddled their oil abundance to construct prosperous and high level social orders. Conversely, nations vigorously reliant upon oil incomes have confronted financial unpredictability and difficulties related with the "asset revile," as changes in oil costs can essentially affect their monetary solidness.

In any case, the persevering through tradition of oil isn't exclusively characterized by its positive commitments. The ecological results of the oil business have been significant, bringing up basic issues about its maintainability. The extraction, transportation, and utilization of oil have been related with a scope of ecological issues, from air contamination to natural surroundings obliteration. Maybe most eminently, the consuming of oil as a fuel source is a significant supporter of ozone depleting substance discharges, which drive environmental change. The outcomes of environmental change, including climbing worldwide temperatures and more continuous and extreme climate occasions, represent an existential danger to the planet and its occupants.

In light of these natural difficulties, there has been a developing familiarity with the need to change away from non-renewable energy sources, including oil. The quest for elective and practical energy sources has picked up speed, with interests in sustainable power advancements like sunlight based, wind, and hydroelectric power. Endeavors to decrease reliance on oil are inspired by ecological worries as well as by worries about energy security, as countries try to differentiate their energy sources and lessen their weakness to oil cost changes.

The oil business itself has confronted huge difficulties lately. Market unpredictability, driven by variables like international struggles, changes in worldwide interest, and the ascent of shale oil creation, has upset customary stock chains and monetary steadiness. This has constrained oil-creating countries and organizations to adjust and enhance their methodologies.

Quite possibly of the most noticeable change in the business has been the ascent of shale oil creation in the US. This mechanical transformation has opened tremendous stores of oil and petroleum gas, prompting a flood in homegrown energy creation. The shale blast has not just changed the U.S. into a key part in the worldwide energy market yet has likewise changed the elements of worldwide oil organic market. The expanded accessibility of oil has, on occasion, prompted a drop in worldwide oil costs, influencing oil-subordinate economies and representing a test to customary oil-delivering countries.

Moreover, the push for cleaner energy sources and the progress to electric vehicles (EVs) can possibly reshape the fate of the oil business. Electric vehicles, fueled by batteries, are viewed as a more manageable option in contrast to conventional gas and diesel vehicles. As innovation propels and the framework for EVs improves, it is conceivable that the interest for oil in the transportation area will decline essentially.

In light of these difficulties, many oil organizations have differentiated their portfolios by putting resources into environmentally friendly power projects and investigating new plans of action. Some have even rebranded themselves as "energy" organizations instead of simply "oil" organizations, mirroring a more extensive spotlight on a scope of fuel sources. While the progress away from oil may not work out more or less by accident, obviously the business is at a basic crossroads, confronting both natural and monetary tensions to adjust.

The persevering through tradition of oil additionally envelops the social and social elements of our reality. The omnipresence of oil has affected our regular routines, from the manner in which we travel to the items we use. It has changed our urban communities, our foundation, and our economies, making a general public that depends on the prepared accessibility of energy. The accommodation and solace that oil has brought to our lives are unquestionable, yet they accompany ecological costs that are turning out to be progressively clear.

Oil has additionally been integral to political talk and worldwide relations. Admittance to oil assets has been a driver of international strategy choices and plays had a significant impact in clashes and partnerships.

The Center East, specifically, has been a locale of worldwide vital importance because of its tremendous oil holds. The US, in its quest for energy security, has been profoundly laced with the governmental issues of the Center East, prompting complex and now and again dubious associations with oil-delivering countries.

Notwithstanding international ramifications, the oil business has confronted examination for its job in molding social and monetary abberations. The dispersion of oil abundance inside oil-delivering nations has been a wellspring of pressure and struggle. The "asset revile" peculiarity has featured the difficulties related with overseeing and appropriating oil incomes, frequently bringing about disparity and defilement inside these countries.

The oil business has likewise impacted the social and imaginative articulations of our general public. Writing, film, and workmanship have frequently wrestled with the subjects of oil, its power, and its ramifications. Books like Upton Sinclair's "Oil!"

9.1 Summarizing the historical significance of the oil industry.

The verifiable meaning of the oil business is immense and complex, traversing hundreds of years and molding the world in significant ways. The business' story starts during the nineteenth 100 years with the penetrating of the principal business oil well in Pennsylvania. This occasion denoted the introduction of an industry that would alter the world.

The development of the oil business matched with the Modern Upset, a time of extraordinary mechanical and monetary change. Oil, at first found as a normally happening asset, turned into an essential driver of the Modern Upset. It gave the energy important to control the steam motors that drove apparatus, transportation, and farming. Subsequently, oil turned into the backbone of a time set apart by phenomenal development and thriving.

The gas powered motor, which depends on oil as a fuel source, further sped up the business' significance in the twentieth 100 years. The ascent of the vehicle business and the broad reception of vehicles as the essential method of individual transportation changed social orders and metropolitan scenes. Suburbanization, with its rambling networks and the appearance of the Pursuit of happiness, was made conceivable by the portability that the car advertised.

Oil's importance expands well past transportation. It fills in as the underpinning of the whole worldwide energy area, from power age to warming. The petrochemical business, as well, intensely depends on oil as a feedstock, creating many items, from plastics to drugs. Oil's effect on our regular routines is obvious in the numerous petrochemical-determined materials we experience.

Geopolitically, oil plays had an essential impact in molding world-wide relations. The mission for control of oil assets has energized clashes and partnerships, frequently at the focal point of worldwide epic showdowns.

The twentieth century saw the ascent of significant oil-delivering countries in the Center East, like Saudi Arabia and Iran, hardening their situation in foreign relations. The international relations of oil have driven international strategy choices and worldwide elements.

Monetarily, oil-rich countries enjoy experienced significant benefits, utilizing their regular assets to assemble prosperous and high level social orders. Nations like the Assembled Middle Easterner Emirates and Norway have really saddled their oil abundance. Interestingly, countries vigorously subject to oil incomes have wrestled with financial instability and the difficulties related with the "asset revile," wherein oil vacillations fundamentally influence monetary dependability.

In any case, the oil business' authentic importance isn't liberated from difficulties. Natural results pose a potential threat, offering basic conversation starters about manageability. The extraction, transporta-tion, and utilization of oil have been connected to a scope of natural issues, from air contamination to environment obliteration. Most quite, the consuming of oil as a fuel source adds to ozone harming substance outflows, driving environmental change and representing an existential danger.

The ecological difficulties have prodded a developing consciousness of the need to change away from petroleum products, including oil. Endeavors to lessen oil reliance are spurred by natural worries as well as by energy security contemplations. Expanding energy sources and diminishing weakness to oil cost changes have become needs for countries and ventures.

The oil business has confronted huge difficulties, from market unpredictability to innovative progressions that have upset customary inventory chains. The ascent of shale oil creation in the US, specifically, has changed the business. This innovative upset opened tremendous stores of oil and petroleum gas, turning the U.S. into a key part in the worldwide energy market.

Besides, the change to electric vehicles (EVs) and the push for cleaner energy sources can possibly reshape the fate of the oil business. EVs, fueled by batteries, are viewed as a more practical option in contrast to customary gas and diesel vehicles. Progressions in innovation and framework for EVs could essentially decrease oil interest in the transportation area.

Oil organizations have answered these difficulties by broadening their portfolios, putting resources into sustainable power projects, and investigating new plans of action. Some have rebranded themselves as "energy" organizations, stressing a more extensive spotlight on different energy sources. While the change away from oil might be slow, the business is at a basic point, confronting natural and monetary tensions to adjust.

The social and social components of oil's persevering through heritage are similarly critical. Oil has turned into a vital piece of our day to day routines, affecting the manner in which we travel, the items we use, and the foundation we depend on. The accommodation and solace it has brought to our lives are unquestionable, however they come at a natural expense.

Strategically, admittance to oil assets has been a driver of international strategy choices and assumed a urgent part in clashes and unions. The Center East has been a district of worldwide vital importance because of its immense oil holds. The US, in its quest for energy security, has been profoundly engaged with the legislative issues of the Center East, prompting complex and in some cases combative associations with oil-creating countries.

The oil business has likewise been examined for its part in molding social and monetary abberations. The dispersion of oil abundance inside oil-delivering nations has frequently been a wellspring of pressure and struggle. The "asset revile" peculiarity features the difficulties related with overseeing and disseminating oil incomes, habitually bringing about disparity and debasement.

Socially, oil's impact has been reflected in writing, film, and workmanship. Various works have wrestled with the topics of oil, its power, and its ramifications. Books as sinclair Upton's "Oil!" and movies like "There Will Be Blood" affect people, networks, and social orders. The social portrayals of oil frequently dive into the moral and moral inquiries it raises.

9.2 Reflection on the industry's impact on society and the environment.

The oil business' effect on society and the climate is a subject of critical concern and discussion. While oil plays had a focal impact in shaping present day culture and driving monetary development, its impacts on the climate have brought up basic issues about supportability and the drawn out results of our reliance on petroleum products.

The social effect of the oil business is diverse and reaches out to different parts of society. On the positive side, oil has been a significant driver of monetary development, work creation, and mechanical advancement. It has supported whole enterprises, like transportation, assembling, and petrochemicals, which have added to higher expectations for everyday comforts and expanded monetary thriving. Oil-rich nations enjoy experienced huge monetary benefits, utilizing their assets to construct progressed social orders with current framework, training, and medical care frameworks.

The oil business has likewise affected cultural designs and urbanization. The far and wide accessibility of oil-controlled transportation, especially the vehicle, has formed the manner in which we live and work. Suburbanization, described by the development of rambling rural areas and the ascent of worker culture, was made conceivable by the versatility presented via vehicles. This change in metropolitan scenes has had significant social and social ramifications.

Besides, the petrochemical business, dependent on oil as a feedstock, has led to a large number of items that are fundamental to present day life. Plastics, manufactured materials, drugs, and incalculable shopper products are gotten from the petrochemical business. The accommodation and solace that these items give have become necessary to our day to day routines.

Be that as it may, these cultural advantages have not come without results. The most squeezing concern is the natural effect of the oil business. The extraction, transportation, and utilization of oil have been connected to different ecological issues, from air contamination and water tainting to living space obliteration and biodiversity misfortune.

Maybe the main natural issue related with the oil business is its commitment to environmental change. The consuming of oil as a fuel source discharges ozone harming substances, especially carbon dioxide (CO_2), into the air. These gases trap heat, prompting a climb in worldwide temperatures and an expansion in the recurrence and seriousness of outrageous climate occasions. Environmental change represents a grave danger to the planet, influencing biological systems, farming, and human networks.

Oil slicks are another notable ecological result of the business. Mishaps and breaks during the extraction, transportation, and handling of oil can bring about devastating spills that hurt oceanic life, harm shorelines, and disturb neighborhood economies. The 2010 Deepwater Skyline oil slick in the Bay of Mexico, for instance, was one of the biggest ecological debacles in U.S. history, making broad harm marine biological systems and waterfront networks.

The natural results of the oil business have powered a developing familiarity with the need to change away from petroleum derivatives and toward cleaner, more economical energy sources. Endeavors to alleviate the effect of oil on the climate are propelled by natural worries as well as by the acknowledgment of the dangers related with environmental change and the earnest need to decrease ozone depleting substance discharges.

In light of these natural difficulties, legislatures, ventures, and common society have sought after different procedures to lessen oil utilization and advance the utilization of sustainable power sources. Interests in wind, sun based, hydroelectric, and thermal power have expanded, prompting a more enhanced energy scene. These endeavors mean to diminish the business' ecological impression and lessen its part in driving environmental change.

One critical improvement in this progress is the ascent of electric vehicles (EVs) as a more manageable option in contrast to customary fuel and diesel vehicles. EVs, fueled by power put away in batteries, produce zero tailpipe emanations. The developing prominence of EVs is viewed as a basic move toward decreasing the transportation area's dependence on oil and its related ecological effects. States and ventures are putting resources into EV foundation and advancements to speed up this change.

The oil business itself has confronted critical difficulties lately. Market unpredictability, driven by variables like international struggles, changes in worldwide interest, and mechanical progressions, has upset customary stock chains and monetary dependability. The ascent of shale oil creation in the US, specifically, has modified the elements of worldwide oil organic market, prompting vacillations in oil costs.

The shale unrest, described by the utilization of pressure driven breaking (deep oil drilling) innovation to extricate oil and flammable gas from shale rock developments, has made the US a central part in the worldwide energy market. The expanded accessibility of oil has on occasion prompted a drop in worldwide oil costs, affecting oil-subordinate economies and representing a test to conventional oil-creating countries.

This moving scene has incited oil organizations to enhance their systems and portfolios. Many have put resources into environmentally friendly power projects and investigated new plans of action that go past customary oil and gas creation. A few organizations have even rebranded themselves as "energy" organizations, mirroring a more extensive spotlight on a scope of fuel sources. These endeavors signal an acknowledgment of the need to adjust to a changing energy scene.

The progress away from oil isn't without its difficulties, including the requirement for foundation improvement, energy capacity arrangements, and changes in shopper conduct. Furthermore, numerous countries remain vigorously reliant upon oil incomes, which can make monetary and political hindrances to such advances. The broadening of fuel sources and the advancement of maintainability have become focal topics in the continuous conversation about the eventual fate of the oil business and its cultural and ecological effect.

The getting through tradition of the oil business is additionally exemplified by its focal job in international elements. Admittance to oil assets has generally been a driver of global struggles and unions. Oil-creating countries have looked to tie down their admittance to these significant assets, prompting complex connections and fights for control on the worldwide stage.

The Center East has been a locale of specific importance in the international relations of oil. Tremendous oil holds in nations like Saudi Arabia, Iran, and Iraq have made the district a point of convergence of worldwide consideration. The US, as the world's biggest purchaser of oil, has had a profound and complex contribution in Center Eastern governmental issues, driven by its quest for energy security and its craving to guarantee a steady progression of oil.

The quest for oil assets has been connected to intercessions, clashes, and wars, frequently revolved around the Center East. These occasions have had expansive results, from the Persian Bay Conflict in the mid 1990s to the continuous struggles in Iraq and Syria. The international relations of oil have been a wellspring of pressure and precariousness, highlighting the business' worldwide importance.

Monetarily, oil-rich countries enjoy delighted in critical benefits, utilizing their assets to construct progressed social orders with high expectations for everyday comforts. The income produced by oil sends out has empowered these countries to put resources into framework, instruction, medical services, and broaden their economies.

Be that as it may, the circulation of oil abundance inside oil-creating nations has frequently been a wellspring of strain and struggle. The "asset revile" peculiarity, as it is known, focuses to the difficulties related with overseeing and appropriating oil incomes. Defilement, imbalance, and political precariousness are normal issues in countries vigorously reliant upon oil incomes. This has prompted inquiries concerning how to best oversee and outfit the monetary advantages of oil abundance for the more extensive populace.

Socially and socially, oil's effect is apparent in various articulations of craftsmanship, writing, and film. These innovative works frequently investigate the subjects of oil's power and its ramifications. Upton Sinclair's book "Oil!" and Paul Thomas Anderson's film "There Will Be Blood" are remarkable models that dig into the business' impact on people, networks, and social orders.

These social portrayals every now and again investigate the moral and moral inquiries raised by the oil business. The quest for abundance, the abuse of regular assets, and the human expenses of industrialization are normal subjects. These works offer a focal point through which society considers its relationship with oil and the more extensive ramifications of its utilization.

9.3 Closing thoughts on the enduring role of oil in our world.

The getting through job of oil in our reality is a subject of extraordinary importance, and it is critical to ponder the intricacies and difficulties related with our proceeded with dependence on this non-renewable energy source. While oil has assumed a focal and extraordinary part in molding the cutting edge world, its impact is presently being rethought considering ecological worries, mechanical headways, and the requirement for a more reasonable energy future.

Oil has been the main thrust behind financial development and industrialization for more than a long period. It controlled the Modern Transformation, empowering the advancement of hardware, transportation, and assembling processes that upset social orders. The overflow and moderateness of oil have supported worldwide economies and added to expanded expectations for everyday comforts. Oil-rich countries have utilized their assets to assemble progressed social orders with present day foundation and top notch administrations.

One of the most striking effects of oil on society has been the multiplication of the auto, which has re-imagined the manner in which individuals live and work. The boundless accessibility of oil-controlled transportation has molded metropolitan scenes and led to suburbanization. This shift has not just affected what we move inside our networks however has likewise meant for social designs and social standards. The accommodation and versatility managed the cost of by the vehicle have become essential to current life.

Besides, oil's significance reaches out to the whole energy area. It fills in as an establishment for power age, warming, and different modern cycles.

The petrochemical business, dependent on oil as a feedstock, delivers a variety of items that are key to our regular routines, from plastics to drugs. The inescapable utilization of petrochemical-determined materials is a demonstration of the compass of the oil business into our cutting edge society.

Nonetheless, the cultural advantages of oil are joined by huge ecological outcomes. The extraction, transportation, and utilization of oil have been connected to a scope of natural issues, including air contamination, water tainting, living space obliteration, and biodiversity misfortune. Maybe most fundamentally, the consuming of oil as a fuel source is a significant supporter of ozone depleting substance emanations, which drive environmental change.

Environmental change represents a grave danger to the planet and its occupants. Climbing worldwide temperatures, dissolving ice covers, and the rising recurrence and seriousness of outrageous climate occasions are among the outcomes of our dependence on petroleum derivatives. The acknowledgment of these dangers has incited a developing familiarity with the need to change away from oil and other carbon-serious energy sources.

Endeavors to decrease oil reliance and advance maintainability are propelled by both biological worries and energy security contemplations. Differentiating energy sources and decreasing weakness to oil value variances are key targets for countries and businesses the same. Sustainable power advancements, for example, sunlight based, wind, and hydroelectric power, have considered expanded speculation and reception to be cleaner choices to oil and other petroleum products.

Perhaps of the most noticeable change in the energy scene is the ascent of electric vehicles (EVs) as a more supportable method of transportation. EVs, controlled by power put away in batteries, produce no tailpipe outflows, offering a harmless to the ecosystem option in contrast to conventional gas and diesel vehicles. Propels in innovation, combined with the advancement of charging framework, are making ready for the boundless reception of EVs.

The oil business has confronted critical difficulties as of late. Market unpredictability, driven by factors like international contentions, changes in worldwide interest, and the ascent of shale oil creation, has disturbed conventional stock chains and financial solidness. The shale upheaval in the US, driven by water powered cracking (deep oil drilling) innovation, has changed the business' scene. The US has turned into a central part in worldwide energy markets, affecting the elements of organic market.

This changing scene has incited oil organizations to broaden their systems and portfolios. Many have put resources into sustainable power projects, investigated new plans of action, and embraced a more extensive spotlight on different energy sources. A few organizations have even rebranded themselves as "energy" organizations instead of stringently "oil" organizations, mirroring the acknowledgment of the need to adjust to a changing energy scene.

While the progress away from oil might be slow, obviously the business is at a basic point. The continuous discussion spins around the harmony between the cultural advantages of oil and the dire need to address its natural and environment related results. This progress presents difficulties connected with framework, energy capacity, and changes in buyer conduct, as well as political and monetary contemplations.

The getting through job of oil is additionally appeared in the mind boggling international affairs of the business. Admittance to oil assets has generally been a driver of worldwide contentions and collusions. Oil-delivering countries have tried to tie down their admittance to these significant assets, prompting mind boggling connections and battles for control on the worldwide stage.

The Center East has been a district of specific significance in the international affairs of oil because of its huge stores. The US, as the world's biggest oil shopper, plays had a huge impact in Center Eastern legislative issues to guarantee a steady progression of oil. This contribution has been set apart by complex, here and there petulant associations with oil-creating countries and a solid spotlight on energy security.

Financially, oil-rich countries enjoy appreciated huge benefits, utilizing their assets to assemble progressed social orders with high expectations for everyday comforts. Notwithstanding, the conveyance of oil abundance inside these nations has frequently been a wellspring of strain and struggle. The "asset revile" peculiarity has enlightened the difficulties related with overseeing and conveying oil incomes. Defilement, imbalance, and political precariousness are normal issues in countries vigorously subject to oil incomes.

Socially and socially, oil's impact is apparent in different types of creative articulation. Writing, film, and craftsmanship have frequently investigated the subjects of oil's power and its ramifications. Books and movies like "Oil!" and "There Will Be Blood" affect people, networks, and social orders. These works give a stage to society to consider the moral and moral inquiries raised by the oil business.

All things being equal, the getting through job of oil in our reality is set apart by its significant and complex effect on society, the climate, and international relations. While it has been a main impetus behind financial development, mechanical headway, and social articulations, it has likewise achieved ecological difficulties and represented a critical danger through its commitment to environmental change.

As we push ahead, it is basic to find some kind of harmony between recognizing the verifiable significance of oil and addressing the squeezing need for a change to more maintainable and naturally mindful energy sources. This change isn't just a mechanical and monetary test however a moral and moral goal. It requires an aggregate exertion from legislatures, businesses, and people to explore the way toward a cleaner, more practical energy future that regards both our cultural necessities and the strength of our planet. The getting through job of oil fills in as a sign of the intricacies and difficulties of this progress, and the basic job we as a whole play in molding the eventual fate of our reality.

www.ingramcontent.com/pod-product-compliance
Lightning Source LLC
LaVergne TN
LVHW051258200726
843510LV00010B/1182